The Early Development of the Aviation Industry

The Early Development of the Aviation Industry: Entrepreneurs of the Sky provides an introduction to the world of the early aviation industry and the business endeavours of the original aviators.

Many of the first pioneers who flew heavier-than-air planes went on to develop considerable industrial concerns. In doing so they exhibited a number of entrepreneurial qualities, which provide useful case studies for those interested in studying how successful entrepreneurs create or develop opportunities at the inception and emergence of high-tech industries. This book looks at the careers of pioneer aviators in the United States, the United Kingdom and France such as A.V. Roe, Thomas Sopwith, Glenn Curtiss and William Boeing. It examines this group of entrepreneurs during the start-up and early development stages of an emerging industry undergoing considerable technological change, and relates this experience to contemporary studies and experiences of entrepreneurship. The book explores what made these men successful in their entrepreneurial endeavours to help promote a better understanding of what makes an entrepreneur and what business and economic conditions are needed to allow such men to be successful.

This book makes a major contribution to our knowledge of the development of the twentieth-century economy and is essential reading for students and academics who are interested in the development of aviation and the nature of entrepreneurial behaviour.

Malcolm Abbott is an Associate Professor of Economics at the Swinburne University of Technology in Melbourne, Australia. The main fields of work in which he has been engaged have been the development of utilities, including airports, energy markets and the manner in which pricing and access to network infrastructure are regulated by government agencies. In addition, he has also worked in other areas of government policy including aviation, rail, dairy industry deregulation, stevedoring and education.

Jill Bamforth is the Academic Director for Work Integrated Learning at the Faculty of Business and Law at Swinburne University, Melbourne, Australia. She lectures and publishes in management and entrepreneurship.

"This book presents a comprehensive overview of early pioneers in the aviation industry and how they achieved many successes against challenging odds. This book is a recommended read for those interested in early aviation. Perhaps there are lessons to be learned that can be applied to modern day aviation entrepreneurship."

– John Wensveen, Ph.D., Vice Provost, Miami Dade College, USA

The Early Development of the Aviation Industry

Entrepreneurs of the Sky

Malcolm Abbott and Jill Bamforth

Routledge
Taylor & Francis Group
LONDON AND NEW YORK

First published 2019
by Routledge
2 Park Square, Milton Park, Abingdon, Oxon OX14 4RN

and by Routledge
605 Third Avenue, New York, NY 10017

First issued in paperback 2020

Routledge is an imprint of the Taylor & Francis Group, an informa business

British Library Cataloguing-in-Publication Data
A catalogue record for this book is available from the British Library

Library of Congress Cataloging-in-Publication Data
A catalog record for this book has been requested

ISBN 13: 978-0-367-72960-8 (pbk)
ISBN 13: 978-1-138-36874-3 (hbk)

Typeset in Times New Roman
by Apex CoVantage, LLC

Contents

Figures

Tables

Part I

Themes and method

1 Introduction

In recent years, considerable interest has arisen about the nature of entrepreneurship and what creates it. This has led to a number of books being written which discourse on the theoretical reasons for entrepreneurship and on the individual achievements of notable entrepreneurs (i.e. Richard Branson, Steve Jobs and Bill Gates). Business academics also are showing a growing interest in the nature of entrepreneurship and there is even some interest in the lives of prominent entrepreneurs by the general reading public. What is lacking, however, is a detailed analysis of how early entrepreneurs in the past successfully negotiated the early stages of development of a highly dynamic industry to establish organisations that went on to shape the future development of that industry. The work's purpose, therefore, is to examine such a group of entrepreneurs, during the start-up and early development stages of an emerging industry, undergoing considerable technological change, and to relate this experience to contemporary studies and experiences of entrepreneurship.

Case studies such as these provide readers and researchers with an opportunity to understand and to further evaluate the impact of events and the dynamics of individual organisations. These types of case studies can also help to advance knowledge and contribute to theoretical developments by bringing to light concepts and ideas that can then be empirically tested (Gummerson, 1991; Nagel, 1961; Yin, 1989; Eisenhardt, 1989). The case studies presented in this book, which concentrate on the early aviation industry, are particularly striking ones and therefore of considerable interest.

Industries undergoing rapid technological change often attract individuals with an entrepreneurial frame of mind, as they offer opportunities to develop ideas and vision. These industries can also generate a degree of excitement that attracts an adventurous type of person. The industry under study in this book is that of the aviation industry, focusing on aircraft manufacturing rather than air travel, which began in the 1900s with the first flights of a heavier-than-air aircraft. In the years up until 1945, this industry saw the start-up of a number of famous aircraft manufacturing companies, assisted by some of the great names in pioneering aviation. Although a number of these men are well known in history, they are generally known more as either pioneers of aviation or as aircraft designers (or both) rather than as entrepreneurs. This is unfortunate because many of them, such as the

Wright brothers, Glenn H. Curtiss, Glenn L. Martin, Louis Blériot, the Farman brothers, Thomas Sopwith, Geoffrey de Havilland, William Boeing, Frederick Handley Page and Robert Blackburn were the creators of huge industrial empires, and the aircraft that their companies produced went on to become household names (for more comprehensive lists of the aviation pioneers/entrepreneurs in the United Kingdom, France and the United States see Tables 1.1, 1.2 and 1.3). These individuals not only had a taste for the excitement of early flight and a great flair in aircraft design, but they also possessed the organisational and entrepreneurial abilities that enabled them to build up considerable industrial concerns. In doing so, some of these individuals found the change in the industry over time too much to handle and struggled to adapt to the demands of managing large enterprises.

Table 1.1 Aviation entrepreneurs in the United Kingdom

Name	*Life*	*Company*	*Founded (taken over) ceased*
Frederick Handley Page	1885–1962	Handley Page Ltd.	1909–1970
Geoffrey de Havilland	1882–1865	de Havilland Aircraft Co. Ltd.	1920–(1960) 1964
Montague Napier	1870–1931	D. Napier & Son Ltd.	1808–(1942)
Frederick Henry Royce	1863–1933	Rolls-Royce Ltd.	1906–
George Holt Thomas	1869–1929	Aircraft Manufacturing Co. Ltd. (Airco)	1912–1920
Alliot Vernon Roe	1877–1958	A.V. Roe & Company (Avro)	1910–(1920) 1963
		Saunders-Roe	1929–1964
Thomas Sopwith	1888–1989	Sopwith Aviation Co.	1912–1920
		Hawker Aircraft. Hawker Siddeley	1920–(1935) 1963 1935–1992
Charles Richard Fairey	1887–1956	Fairey Aviation Co. Ltd.	1915–(1960) 1977
Oswald Short	1883–1969	Short Brothers PLC	1897–(1989)
Eustace Short	1875–1932	Short Brothers PLC	1897–(1989)
Robert Blackburn	1885–1955	Blackburn Aircraft Limited	1914–1960
George White	1854–1916	British and Colonial Aeroplane Co.	1910–(1959) 1966
Nevil Shute Norway	1899–1960	Airspeed Ltd.	1931–(1940) 1951
Sam Saunders	1857–1933	S.E. Saunders (Saunders-Roe)	1896–1929
Noel Pemberton-Billing	1881–1948	Pemberton-Billing Ltd. (Supermarine Aviation Works Ltd.)	1913–(1928) 1960
Ernest Petters	1873–1954	Westland Aircraft Ltd.	1915–2001
Hugh Burroughes	1884–1985	Gloucestershire Aircraft Co. Ltd. (Gloster)	1917–(1934) 1963
John Davenport Siddeley	1866–1953	Armstrong Siddeley	1919–(1935) 1960
James Martin	1893–1981	Martin Baker Aircraft Co.	1934–

Table 1.2 Aviation entrepreneurs in the United States

Name	*Life span*	*Company*	*Founded (taken over) ceased*
Wilbur Wright	1867–1912	Wright Co. 1909–1929	1909–1929
Orville Wright	1871–1948	Wright Co. 1909–1929	1909–1929
Glenn H. Curtiss	1878–1930	Curtiss Aeroplane and Motor Co.	1909–1929
Glenn L. Martin	1886–1955	Glenn L. Martin Co., Wright-Martin	1912–1961
Frederick B. Rentschler	1887–1956	Pratt & Whitney Aircraft Co.	1925–
Allan Loughead	1889–1969	Alco Hydro-Aeroplane Co., Loughead Aircraft Manufacturing Co.	1912–1920
William E. Boeing	1881–1956	Boeing Co.	1916–
Jack Northrop	1895–1981	Avion Corp., Northrop Corp.	1939–1994
Donald Wills Douglas	1892–1981	Douglas Aircraft Co.	1921–(1967)
Larry Bell	1894–1956	Bell Aircraft	1935–
Gerard Freebairn Vultee	1900–1938	Vultee	1939–(1943)
Vance Breese	1904–1973	Breese Aircraft Co., Vultee	1939–(1943)
Sherman Fairchild	1896–1971	Fairchild Aircraft	1925–(2003)
James Smith McDonnell	1899–1980	McDonnell Aircraft Corp.	1939–1967
Reuben Hollis Fleet	1887–1975	Consolidated Aircraft	1923–1943
James Kindlebeger	1895–1962	North American Aviation	1928–(1967)
Leroy Grumman	1895–1982	Grumman Aeronautical Engineering	1929–1994
Chance M. Vought	1890–1930	Lewis and Vought Corp., Chance Vought Corp.	1917–1962
Alexander P. de Seversky	1894–1974	Seversky Aero Corp.	1931–1939
Grover Cleveland Loening	1888–1976	Loening Aeronautical Engineering Corp., Grover Loening Aircraft Co.	1917–1932
Giuseppe Mario Bellanca	1886–1960	Bellanca Aircraft Co.	1927–1954

Table 1.3 Aviation entrepreneurs in France

Name	*Life span*	*Company*	*Founded (taken over) ceased*
Gabriel Voisin	1880–1973	Voisin, Société Anonyme des Aéroplanes G. Voisin	1905–1918
Charles Voisin	1882–1912	Voisin	1905–1918
Henri Farman	1874–1958	Farman Aviation Works	1908–1936

(*Continued*)

Table 1.3 (Continued)

Name	*Life span*	*Company*	*Founded (taken over) ceased*
Louis Blériot	1872–1936	Blériot Aéronautique	1908–1936
Lois Breguet	1880–1955	Breguet Aviation	1911–1971
Edouard de Nieport	1875–1911	Société Anonyme Des Établissements Nieuport	1908–1937
Louis Bechereau	1880–1970	Société de Production des Aéroplanes Deperdussin	1911–1921
Raymond Saulnier	1881–1964	Morane-Saulnier	1911–1966
Aldophe Bernard	1881–1955	Avions Bernard	1917–1935
Émile Dewoitine	1879–1979	Constructions Aéronautiques Émile Dewoitine	1920–1927
Marcel Bloch (Dassault)	1892–1986	Avions Marcel Bloch; Dassault Aviation	1929–
Felix Amiot	1897–1974	Avions Amiot	1916–1945
Pierre-Georges Latécoère	1883–1943	Groupe Latécoère	1917–
Henry Potez	1891–1981	Aviation Potez	1919–1967
Fernand Liore	1874–1966	Lioré-et-Olivier	1912–1937
Lawrence Santoni	1877–1928	CAMS	1920–1937
Etienne Romano	1899–1966	Romano	1929–1936

Others, however, were able to negotiate the changes well, successfully remaining in the industry long into the twentieth century, and in the process becoming substantial captains of industry. One example, for instance, is that of the British flyer Geoffrey de Havilland, who was one of the very first aviators in the United Kingdom, initially flying an aircraft he built in 1909 (see Figure 1.1). He and his companies survived the complex World War I and World War II industry environments and his company became an important designer of jet aircraft in the post-war years (see for an example Figure 1.2). Understanding why some of the early aviation entrepreneurs were able to make the transition from aviation pioneer to industrial manager while others did not will be an important theme of this study.

Looking back on the careers of these entrepreneurs it is easy to forget that most of them began their companies in small workshops and garages tinkering with crude machines that they hoped would enable them to experience flight. In doing so they are to some degree an earlier equivalence of the computer enthusiasts of the 1970s and 1980s, who developed early computer hardware and software that went on to shape the later-day IT industry, and subsequently all of our lives. Insights can be found by examining the similarities between the small-scale workshop origins of the aviation industry and the early stages of the personal computer revolution of the 1970s and early 1980s, when individual entrepreneurs on shoestring budgets worked out of garages and other modest locations, developing path-breaking concepts and technologies that helped to create the information

Figure 1.1 The first aircraft designed by Geoffrey de Havilland (1909)

Figure 1.2 Preserved de Havilland Vampire T11 displays at the Cotswold Air Show at Cotswold Airport, Kemble, Gloucestershire, England

revolution. The pioneering period of aircraft manufacturing provides a rich illustrative history of the growth of a new industry, which involves the creation of a number of small firms, financed from individual, family or local resources, all edging their way into the unknown.

In observing the similarities between early aviators and computer enthusiasts, there are some differences. Perhaps the main one was that although the latter were prepared to risk their careers and finances on their IT obsessions, the pioneer flyers were prepared to also risk their lives to achieve their personal goals. To some degree, however, this is not really a difference in general attitude, but instead a difference in the degree of intensity of the risks they were prepared to take. The early aviation and IT pioneers were both enthusiasts of emerging technologies and fascinated by the possibilities for the future they promised, it was just that the aviators also took on a rather more personal, physical risk as well.

Not all of these enthusiasts in these two industries went on to found and develop great industrial concerns, but in the case of aviation, a number did so. To a degree, they formed these industrial concerns to support and develop their enthusiasms, but often these companies developed beyond that into large-scale enterprises. Understanding what made these men successful in their entrepreneurial endeavours will go a long way to helping to promote a better understanding of what makes an entrepreneur, and what business and economic conditions are needed to allow such men to be successful.

To begin with, it is important to be mindful of what exactly is an entrepreneur. Entrepreneurs are said to display a number of characteristics that differentiate them from conventional business managers. Entrepreneurs generally focus on the achievement of a long-term vision and use the resources available to them (networks, knowledge, experience and partnerships) to form and iteratively shape a path to that future. Non-entrepreneurs who manage businesses, in contrast, generally focus on identifying the causes of predictable outcomes that they might be expected to face and then try to manage their environment using existing frameworks and resources to deliver those outcomes. The early pioneers of aviation who went on to design aircraft and build up companies to support their endeavours possessed many of the characteristics of what we generally regard as being entrepreneurs. Often these characteristics spurred them on to start-up new enterprises but also, ultimately, it meant they became bored when their enterprises grew to a substantial size, and they were forced into the role of being large-scale industrial managers.

Inter-linked to the endeavours of these early aviators/entrepreneurs is the technological context in which they worked. To understand the characteristics that early aviators/entrepreneurs had it is important to appreciate the waves of technological advancement that occurred in the aviation industry starting in 1903. In heavier-than-air aviation, the product had to meet exacting standards of technical quality, as the inherent dangers associated with the emerging technology were ever present. This put considerable development pressures on these people.

It is also important to understand the nature of consumer demand for aviation (coming as it does from a limited number of customers). These customers consisted of demands from governments for military aircraft, commercial aviation companies, a limited private market and in some cases, exports. Aircraft manufacture in its earliest years tended to be concentrated in a limited number of countries, a factor that remains today. The nature and development of the market is, therefore, a consistent theme in aviation and aircraft aerospace development. The study of the initiation of this industry promises to provide a number of fascinating insights into the nature of the entrepreneurial individual, both then and now.

The general purpose of this study is to examine the nature of a group of individuals who worked in aviation in the years 1903 to 1945, the period during which the aircraft manufacturing industry emerged. It will concentrate on explaining the work of these individuals in terms of their entrepreneurial abilities, rather than their activities as pioneer aviators or aeronautical designers and engineers (although descriptions of these aspects of their lives will be provided, especially to the degree it reflects on their entrepreneurial activities). In addition, this study will look at the nature of the industry in which they worked to see what role the dynamic nature of the industry had in both attracting men of this calibre and providing them with the opportunities to develop their entrepreneurial skills and their organisations.

The first aim of the study, therefore, is to give an explanation of the structure of the aviation industry in the period 1903 to 1945, the nature of technological change and the general entrepreneurial context in which these men worked. The industry is widely believed to have begun in 1903 when the Wright brothers at Kitty Hawk in the United States made their first flight. This event heralded in a period of considerable expansion. The first French aircraft flight took place in 1906 and the first British flight in 1908, and the first orders for military aircraft took place soon after that. Military demand during World War I (1914–1918) helped to transform the industry from a small-scale, workshop-based industry into a vast and complex industry, although immediately after the end of the war, the industry went through a painful contraction.

Although the average number of post-World War I orders for aircraft were initially very limited, production and sales increased from 1923 onwards (Fearon, 1978 p. 237). A brief contraction took place in the early 1930s and then the industry began to grow again, pushed on first by the development of commercial passenger aircraft and later, after 1935, by rearmament. This increased demand helped to recreate a massive industry in the United Kingdom and France, and later still in the United States, employing hundreds of thousands of workers. During these years, a second wave of technological innovation in the industry occurred, with the introduction of metal construction, which created stronger aircraft and greater speeds, which in turn required greater technical effort. As the design and production of aircraft became more complex, more teamwork, a broader distribution of skills and a more sophisticated organisational structure became required. The design and manufacture of aircraft then moved on from a reliance on intuitive design skills to more systematic and scientific approaches being used. These technological changes demanded greater economies of scale in the industry, while simultaneously increasing considerably the costs of entry into the industry. These factors eventually put pressure on firms to combine in order to reduce average production costs. Technological growth within the industry provided considerable scope for young men to develop their skills, both engineering and entrepreneurial. The novelist Nevil Shute Norway, who worked as a young man in the industry in the 1920s and 1930s, commented in his autobiography on this aspect:

> It was exciting to be in aviation in those days, because development went at such a pace. . . . In aviation at that time there were opportunities on every side for those who had the wit to take them.
>
> (1954, pp. 28–29)

This general climate of change provides a reader with an interesting context in which to study industrial development. In the case of aviation, however, although there are history books on the pioneers, there is limited discussion of those who were entrepreneurs, their work and how it led to the establishment of organisations and technologies that established the DNA of the current aviation industry.

The second aim of the study is to link the findings from the study of this early history to modern-day entrepreneurship. For example, the book will explore the interconnection between twentieth-century history, types of entrepreneurship, behaviours and psychological makeup using the early aviation industry as a case study. Early entrepreneurs in aviation will be compared to modern-day entrepreneurs in high-tech, emerging industries, to compare and contrast drivers and the patterns of entrepreneurship in different eras. In doing so, it will be possible to see how a dynamic, emerging market operates where established frameworks of knowledge do not yet exist. The study will explore the important role of communities of practice as alternative learning arenas in emergent high-tech development until government funding kicks in. This community of practice aspect would have appeal in contemporary emerging fields such as drone technology.

The third aim of the study will be to provide portraits of some of the most important aviation entrepreneurs in the industry during the period 1903 to 1945. The founders of the early aviation firms were often well-trained engineers, many of whom could be described as entrepreneurs (such as the Wright brothers, Louis Blériot, Glenn H. Curtiss, the Short brothers, Robert Blackburn, A.V. Roe, Geoffrey de Havilland and Thomas Sopwith; see Tables 1.1, 1.2 and 1.3). Although these men have been written about in the past, this work has mainly concentrated on their lives as aviators, designers and engineers rather than as entrepreneurs (see for instance Barnes, 1976; Bramson, 1990; Nockolds, 1950). This is unfortunate as they are a fascinating group of businessmen in their own right and in some cases are amongst the United States, France and the United Kingdom's most important industrial leaders in the twentieth century.

The fourth aim will be to look at the personal characteristics these men had, and how they used their skills to create companies that enabled them to pursue their personal goals and vision. In each case, these men had to balance the need to create companies that enabled them to develop innovative aircraft designs embodying new technologies with soundly run company structures that would attract outside investors. This balancing act was a perpetual problem for the early aviation entrepreneurs and their companies, and was tackled in a variety of interesting ways. Explanations of how they achieved this balance will be a central focus of the book.

The final aim is to look at the way in which aviation entrepreneurs interacted with the industry in other countries (e.g. the interactions of United States, the United Kingdom and France, with countries like Germany, Japan and Italy) to pick up and adapt innovations to meet the needs of their target markets.

In undertaking this work the authors have decided to limit their study to three main countries: the United States, France and the United Kingdom. Although it is true that a number of other countries (e.g. Russia, Germany, Italy and Japan) had similar aviation pioneers, it was decided to limit the study to provide a deeper analysis and make it manageable. Looking at Tables 1.1, 1.2 and 1.3 reveals that not all important aviator entrepreneurs in the three focused countries were included in this study. Again, this is not because those omitted were unimportant, but because including all would lead to a shallower, less clear analysis. It is hoped that in the future, researchers will take a renewed interest in these entrepreneurs of the past, many of whom have largely disappeared from public consciousness, even though as individuals they offer much for us to learn.

References

Barnes, C. H. (1976). *Handley page Aircraft Since 1907*. London: Pavilion Books.

Bramson, A. (1990). *Pure Luck: The Authorized Biography of Sir Thomas Sopwith*. Wellingborough, UK: Patrick Stephens.

Eisenhardt, K. M. (1989). Building theories from case study research. *Academy of Management Review*, 14(4), pp. 532–550.

Fearon, P. (1978). The vicissitudes of a British aircraft company: Handley page Ltd. between the wars. *Business History*, 20, pp. 63–86.

Gummerson, E. (1991). *Qualitative Methods in Management Research*. Newbury Park, CA: Sage.

Nagel, E. (1961). *The Structure of Science: Problems in the Logic of Scientific Exploration*. New York, NY: Harcourt, Brace and World.

Nockolds, H. (1950). *The Magic of a Name*. 2nd ed. London: G. T. Foulis.

Norway, N. S. (1954). *Slide Rule: The Autobiography of an Engineer*. London: Heinemann.

Yin, R. (1989). *Case Study Research: Design and Methods*. Newbury Park, CA: Sage.

2 Entrepreneurship, its origins and recent experience

2.1 Introduction

In the period from 1903 to 1945, it could be said that the aviation industry was 'an emerging' industry. Emerging industries are ones in the earliest stage of development, and they are characterised by periods of unpredictability, where it is difficult for managers and investors to analyse cause and effect patterns in the business environment. This makes it difficult to predict the future and therefore to plan in a methodical way what to do next. This type of environment, which can take between 2 and 50 years to stabilise, favours a sense-making business approach that is sensitive and responsive to changes and evolving opportunities (Forbes and Kirsch, 2011).

Seeing what entrepreneurial patterns there are in emerging industries are important if we are going to be able to understand the very basic nature of the entrepreneurial process. This chapter's purpose, therefore, is to give a brief overview of how successful entrepreneurs survive and thrive in emerging industries. In doing so, early aviation industry examples will be used to highlight common patterns that transcend time and to provide insight into the factors that are likely to contribute to the transition of organisations from start-up to dominant players.

2.2 Entrepreneurship in emerging industries

Organisations build market economies and capitalist societies. In established industries, new entrant companies follow the paths set by their predecessors and survive by catering to an existing niche. In emerging industries, new ventures are the first of their kind and they depart from the established order in a coherent and self-sustaining way. In emerging industries, entrepreneurs, in creating new ventures, collaborate and compete with other entrepreneurs to construct organisations, then populations and finally communities (Aldrich and Martinez, 2003). Individual entrepreneurial activity, therefore, shapes industry dynamics and new knowledge development, which affects how market decisions are made and their outcomes.

Founder responses to established versus emerging industry dynamics

Established markets are ordered in ways that favour decision-makers who make decisions based on predictable cause and effect patterns. These individuals are known as causation decision-makers. In these markets, competition is generally in play and well connected. Efficient business operators who can correctly predict aspects of an uncertain future can gain control and establish a competitive advantage. Systematic information gathering and reasoning helps to select optimal strategies that exploit preexisting knowledge, reducing risk and maximising potential returns. Careers in these ordered environments are structured, planned and linear with infrequent changes in career fields. Learning predominantly comes through attendance at prestigious learning institutions, which extoll tried and tested approaches. These educational institutions often provide access to exclusive networks of industry influencers and decision-makers, enabling companies to develop, refine and exploit their specialist knowledge, networks and access to resources to become dominant industry players. Ordered environments, therefore, favour individuals and businesses that are strongly administrative in nature.

In contrast, emerging markets tend to be disordered and unpredictable. Cause and effect patterns are not easily identified, so pattern recognition is important. In this climate, creating a company in a market that does not yet exist involves being able to understand how to make critical decisions in the absence of preexistent goals, by tapping into communities of practice at the forefront of learning about the emerging technology associated with the new industry. Knowledge is emerging and information is gathered through experimental and iterative learning techniques. Action is, therefore, favoured over planning as it enables those undertaking it to learn as they go. In these circumstances there is a focus on affordable losses, rather than expected returns, and creating options in the future over maximising returns in the present. The types of individuals that thrive in these circumstances tend to be concerned with the effects of their decisions and are therefore known as effectuation decision-makers.

In disordered, unpredictable markets, strategic alliances and pre-commitments from stakeholders reduce or eliminate uncertainty and often can erect barriers to entry. This encourages and supports the development of communities of practice, made up of people who work together to establish, shape and develop the emerging knowledge. Careers can be transitory or spiral and are characterised by frequent changes of fields, organisations and jobs. This allows for individuals to enjoy variety, independence and the exploration of new activities closely related to previous ones. These environments favour individuals and businesses that are entrepreneurial.

In essence, the skills needed to start and run a business in an established industry differ significantly from those required to start and run a business in an emerging market. This impacts the success rates of entrepreneurial start-ups in emerging industries as the industry becomes more established.

Entrepreneurial learning leads to the development of knowledge that is needed for starting and managing new ventures. Politis' framework of entrepreneurial learning suggests that entrepreneurial knowledge is often derived from prior career experience, which can raise the effectiveness of entrepreneurs at recognising and acting on entrepreneurial opportunities and coping with the liabilities associated with newness. Three factors affect the conversion of experience to knowledge: the entrepreneur's predominant logic of reasoning, career orientation and the outcome of previous events. In this chapter, we use Sarasvathy's work on entrepreneurs' decision-making to examine their predominant logic of reasoning and its effect on opportunity exploitation (Sarasvathy, 2009). In this way, we expose underlying beliefs about the future and how to engage with it. Sarasvathy contrasts two decision-making approaches (causation and effectuation) along seven logic dimensions that influence opportunity exploitation: the aspects assumed as givens, decision-making criteria, competencies employed, context viewed as relevant, the nature of unknowns, underlying logic for engagement and desired outcomes.

The first decision-making approach is causation thinking. Causation thinking focuses on how to create desired future outcomes in an assumed static, linear and independent environment, where uncertain future consequences can be avoided through prediction and control, using competitive strategies. Systematic information gathering and reasoning helps to select optimal strategies that exploit preexisting knowledge to reduce risk and maximise potential returns. Competition is often in play, and well-connected efficient business operations deliver competitive advantages. Specialist knowledge can be developed, refined and exploited to enable a firm or individual to become a dominant player, able to exploit existing networks of friends and business acquaintances with mutual needs and interests to access resources. Causation thinking favours ordered environments, established markets and learning from established institutions such as universities that can support the development of powerful connections and best practice approaches to uncover cause and effect relationships and to establish a linear career.

In contrast, effectuation thinking assumes the future does not need to be predicted as it can be created through partnerships, alliances and cooperative strategies in a dynamic, well-connected industry ecosystem. This approach favours making the most of the means available to the individual or firm. Communities of practice, which informally bind together groups of people by shared expertise and passion for a joint enterprise, are of significant importance as they bring together individuals seeking to design the future by exploiting contingencies in the present environment. This approach favours the act, sense and response approaches needed to identify patterns in unordered environments such as in emerging markets where decisions must be made in the absence of preexisting goals. Knowledge is emerging and not established, favouring action over planning. There is a focus on affordable loss rather than on expected returns and creating options in the future over maximising returns in the present. Communities of practice allow members to develop strategic alliances, to build and exchange knowledge, and to reduce or eliminate uncertainty, while also erecting barriers to entry for non-practitioners. Thus, practitioners in communities of practice work together to establish, shape and develop the emerging knowledge. As emerging industries result in individuals with frequent changes of fields, organisations and jobs, this enhances learning and knowledge creation by delivering variety, independence and the exploration of new activities closely related to previous ones.

In summary, unordered, emerging new markets demand different types of behavioural responses and learning approaches than for ordered, established markets. Effectuation reasoning suits the founding and early development stages of emerging industries, when learning and knowledge are emerging, consumer tastes are ambiguous and there are limited technical, academic, industrial or government resources available. Effectuation logic allows individuals to work with others within affordable loss and acceptable risk parameters to exploit contingencies to co-create a future through alliances, cooperative strategies and communities of practice. Causation reasoning suits developed, stable industries where expertise in existing knowledge and being part of dominant networks of industry friends and business acquaintances delivers competitive advantage. Causation logic taps into and exploits existing patterns using personal knowledge to choose between strategies and networks that enable control of a market, build market share and deliver an expected return. Examining entrepreneurial careers and the contexts in which these entrepreneurs started their businesses provides insights into why some were more successful than others in their ventures.

The lifecycle of emerging industries sparked by a radical innovation

At the start of an industry's life, sparked by radical innovation, there is a fermentation period. During this period there is a gap in codified knowledge about the basic nature of the industry's technology and so the pioneering founders must work either independently or together to create a new knowledge base. This knowledge base may be co-created with other founders of organisations or through the transfer of knowledge and evidence from related industries. If the knowledge transfer is limited by tensions between industries and organisations, entrepreneurial founders and their organisations must frame the unknown in such a way that it becomes believable (Aldrich and Fiol, 1994). The growth and credibility of a new industry's community is, therefore, dependent on entrepreneurs and their organisations carving out a legitimate niche and delivering legitimate products, services and knowledge to a committed user base, under conditions of ignorance, uncertainty and low knowledge levels.

Carving out a legitimate niche means the emerging industry must build credibility with a range of stakeholders: e.g. the general public, key opinion leaders, government officials and others operating within or who are affected by the emerging industry. New designs, routines and competencies must be developed and standardised in order to win over a committed user base. In the initial stages of the industry, a lack of standardisation and legislation in these areas attracts new entrepreneurs who flood into the emerging industry, vying to establish the dominant design. However, if the emerging industry is unable to collectively agree on taken-for-granted standards and designs for the development of products and/or services, the industry will fail to establish legitimacy and will struggle to grow.

The era of fermentation comes to an end when participants settle on a dominant design (Anderson and Tushman, 1990). The subsequent era of incremental change brings a rapid convergence towards an accepted design, which new ventures easily adopt rather than seek further fundamental innovation. When a dominant design does emerge, disbandings of organisations increase, entry of new organisations slows and incumbent firms are advantaged as new entrants have less chance of changing the standards of the population.

The importance of legitimacy and credibility in growing an emerging industry

The credibility and legitimacy of an industry are necessary to achieve a critical mass of consumers and support. To achieve this, entrepreneurs must develop effective routines and competencies, and collectively agree standards and designs for product and service development that are accepted by a range of stakeholders. Effective trust building and reliability-enhancing strategies are necessary to establish trust in, and acceptance of, the emerging industry by public and government, but this process is not straightforward. Innovative entrepreneurs must compromise some autonomy in order to enter cooperative relationships, while at the same time competing in broader environments with more actors.

Important to this process is that of technological innovation. Technological innovation often typically involves a cumulative series of inter-related acts of selection, variation and retention that might lead to commercial applications (Garud, 1994). Development of core knowledge, products and services is often part of a larger symbiotic system of components coming from changes in other industries. These industry changes may be stable in one industry area and dynamic in another. Entrepreneurs' strength lies in their ability to capitalise more rapidly on technical advances than bureaucratic organisations and populations to make innovation relevant to the potential user base. This relevance delivers the critical mass of users needed to fund the growth of the industry.

Cooperation versus competition

The nature of the relations between organisations and between industries, whether cooperating or competing, affects resource distribution in the environment and the terms on which they are available to entrepreneurs. In established industries, the lack of cognitive legitimacy of emerging organisations can be both an advantage and a disadvantage. Leading established organisations may overlook emerging organisations and perceive them as a non-threat, allowing the emerging organisation to develop without its legitimacy being undermined by rumours and information suppression or inaccurate dissemination.

In an emerging market, carving out new information can be a dangerous exercise for entrepreneurs. Working with others is often necessary to allow a fine graining of knowledge transfer and joint problem solving, but too much divulgence and dependence on others can threaten an organisation's survivability. Partners may steal knowledge from others to develop more efficient organisations that better meet customer and stakeholder needs. Trust is key and generally emerges from collective interaction and understanding developed over time.

The level of competition between organisations can depend on the size of the population of entrepreneurial organisations and on the degree of similarity, or niche overlap, between these organisations. Until industry standards are set, competition between companies tends to be high. Disagreement on a dominant design increases the number of entrepreneurs entering the marketplace and limits those leaving the marketplace, as firms struggle to have their designs accepted. This process stimulates the development of new ways of doing things or leads to the extensive modification of old ones. Founders of firms compete to have their approach taken for granted, appealing to potential customers, investors and others to accept their approaches (Aldrich, 1999).

Confusion undermines the cognitive legitimacy of the industry as a whole and, therefore, there is pressure on the early founders to create a population standard that delivers innovations that can be imitated by others. Imitable designs attract others into the market, allowing it to grow and become more established. Increased competition generates more collective action and more competition. An expanding market increases entrant survivability, but easy imitability means the market becomes crowded with equally competent rivals. In these circumstances, survival of a firm can become contingent on fairly small differences between ventures, and earlier entrants may be outmanoeuvred by the later entrants. Thus, there is a strong incentive for early founders to cooperate on stabilising conditions in the industry by creating standards that favour them. In markets where there are several different, incompatible designs competing for dominance, consumers become confused and hold back on purchasing until a clear winner emerges.

It is possible that extreme uniqueness can disadvantage an individual firm. An individual firm having a monopoly over a technological innovation does not prevent success by others in the new, emerging markets. Independent firms may choose to exchange and even diffuse their technologies (in contrast to protecting them through patent enforcement) to create product groups that can be packaged and sold. The existence of imitators and complementary products can lead to an increase in the likelihood of a particular innovation being adopted and diffusing, which in turn can lead to the creation of economies of scale, which in turn can lead to lower average unit costs and higher profits (Funk and Methe, 2001, p. 591).

Neither extreme imitability nor extreme uniqueness favours a firm's survival within new populations. The net effect of imitability is contingent on whether a population has converged on a dominant design and on the relative balance between underlying market growth and the existence of new entries and exits from that population.

Building credibility and legitimacy requires standards. The process of achieving standards is, therefore, a socio-cultural one, rather than being purely technologically driven (Rosenkopf and Tushman, 1994). The creation of industry standards may increase the potential for the emergence of synergy amongst participants in an informal network of organisations and increase the survival capabilities of the population in an aggregate. Standards, however, also play a role in discouraging entry and eliminating some organisations within the industry.

2.3 Institutional enablers in emerging industries

In promoting the development of emerging industries, a number of institutional enablers can exist. To begin with, the broader community can influence the rate at which a new industry grows by affecting knowledge diffusion about a new activity and the extent to which it is publicly or officially accepted. Industry founders who pursue effective trust building and reliability-enhancing strategies can help to establish a reputation vis-à-vis other industries, laying the groundwork for attaining cognitive legitimacy at the community level and enhancing the survivability of the emerging industry. Within the community, a number of organisations can act as important enablers or limiters in the development of an emerging industry (e.g. trade associations, educational institutions, government organisations and the media). Some of the more important of these are described next.

Trade associations

In some kinds of technology regimes, new innovations tend to be pursued by a handful of actors, working independently but in parallel. These people can come to know about one another and their work through personal interaction and through travelling in similar technical/social circles, which might include their attendance at the same industry conferences and technical committee meetings. These players can generate social networks that, collected together, result in population-level collective action (Van de Ven, 1991). These informal inter-firm relations can develop into more formalised strategic alliances and the formal creation of standard-setting bodies, such as trade associations (Powell, 1990).

Trade associations are collective interest associations, often started by industry champions who fund start-up costs, with the goal to advance the interests of an organisational population by formulating and promoting product/process standards via trade committees and publishing trade journals. Trade associations build understanding in the general population by facilitating cross-industry learning and cooperation, enhancing the emerging industry's learning and legitimacy. Their minimalistic organisational structure allows them to operate with low overhead and to quickly adapt to changing conditions.

Educational institutions and communities of practice

In addition, educational institutions can create and spread knowledge about dominant competencies, putting resources into the hands of potential founders and increasing the diversity of organisational communities. As technologies stabilise, identifiable jobs emerge for which educational institutions provide training. Courses and degrees train basic competences that formalise and centralise information. Educational institutions, therefore, stimulate growth by spreading generalisable competencies and skills to nascent entrepreneurs. Communities of practice of emerging practitioners in the new industry often form to fill the knowledge gap between what existing educational institutions can provide and what knowledge is needed to exploit opportunities in the dynamic emerging market.

Government support and regulation

Further institutional support can come from government organisations which can support and assist emerging communities, creating stability and accelerating the legitimacy of the new population through research support and through the enactment and enforcement of new laws, regulations and standards. Governments can also help to structure the inter-organisational environment to encourage mutual dependence that benefits both firms and populations.

The media

Finally, the media often plays an important role in disseminating information within and between populations (the public, investment community, military, inventors, entrepreneurs, advocates, educational institutions, hobbyists, etc.), stimulating potential entrepreneurs to perceive opportunities and to combine old resources in new ways or to recognise and realise opportunities in existing populations.

In conclusion, therefore, for an emerging market to survive it must build a viable customer base by establishing legitimacy and credibility. To do this requires setting collectively agreed upon standards and designs that provide a clear product and service development path that reduces risk to customers. In the early start and fermentation stages of an industry, multiple entrepreneurs compete and cooperate to co-create the standards that shape both the present and the future. Trade associations collect like-minded organisations together to promote standards and developments. In addition, educational institutions can create and spread knowledge about dominant competencies and put knowledge resources into the hands of potential founders. Government organisations can also create stability in the emerging community and accelerate its legitimacy through support and assistance by incorporating emerging regulations and standards into laws. Such an ecosystem links entrepreneurs and their organisations to other industry populations and the broader community.

2.4 The entrepreneurs' role in developing early aviation, 1903 to 1945

In taking the basic notions of the general characteristics of an entrepreneur in an emerging industry, already described, it is now possible to see how these notions can apply to the aviation industry in its earliest years. To do this it is necessary to understand the nature of the early development of the aviation industry and how entrepreneurs and their organisations worked with related industries and the community to build the credibility and legitimacy of heavier-than-air aircraft. Table 2.1 provides an overview of the dominant aircraft design types

Table 2.1 Dominant aircraft types by design and country of origin, 1910–1945

	1900–1909	*1910–1919*	*1920–1929*	*1930–1939*	*1940–1949*	*Total*
France	1 T/R	38 F,M	10 F,R	18 B	20 F,M,R	87
Germany	0	86 F,R	1	20 M,B,F	29 M	136
Nazi Germany	0	5 B, R/M	0	28 F,B,M	97 F,B,M	130
Italy	0	7 B,F	2 F	13 F,M,B	17 F,M,B	39
Japan	0	0	0	10 F,B,M	78 F,B,M	88
Russia	0	0	4 F, R/B	8 F,B,M	41 F	53
United Kingdom	2 E	54 M,F	12 F,O	39 B,F,M	56 F,B,M	163
United States	1 TD	11 T/M	20 F/B	46 F,B,M	172 F,B,M	250

Source: Developed by authors from www.militaryfactory.com/aircraft/aircraft-by-decade.asp

Key:
E = Experimental
TD = Technical demonstrator
T = Trainer
R = Reconnaissance
F = Fighter
B = Bomber
M = Multi-modal design for multiple uses
O = Other

and country of origin during the period 1910 to 1949. From 1900 to 1909 the focus was on the building of demonstration and reconnaissance aircraft. The immediate pre-war period and World War I period from 1910 to 1919 saw a shift in emphasis to military fighters, bombers and multi-modal planes, i.e. planes that could carry out multiple functions. From 1920 to 1929 (the interwar years), fighters remained the primary production plane, followed by bombers. During this period, viable transport/passenger planes also began to appear. The preparation period before World War II (1930 to 1939) saw fighters, bombers and multi-modal planes dominating production. Domestic long-range passenger and transport plane production continued to grow slowly during this period. Unsurprisingly, the 1940 to 1949 (World War II and immediate post-war years) period continued with a preference for fighters and bombers. The interest in transport planes also grew, as did an increase in the development of interceptor and jet planes (see Tables 2.2 and 2.3). In the following decades, the focus moved to primarily fighter and transport planes. This section discusses this shift in focus and the interplay of various population groups, e.g. the public, the military and the media.

Table 2.2 Aircraft production in selected nations, 1940 to 1944

	Fighters	*Carrier*	*Recon*	*Heavy bombers*	*Other bombers*	*Trainer/ transport*	*Totals*
United Kingdom	43,000	7,400	3,300	12,000	18,000	47,800	131,500
USA	48,000	53,000	3,200	36,000	24,000	139,500	303,700
Germany	59,500+		c.5,200	c.1,200	32,000+	4,800	111,800

Source: Goldberg, Equipment and services (1955, pp. 350–52); British totals from Postan, *British War Production* 484, cited in Lund (1998, p. 99)

Table 2.3 Designs ordered into production

	9/39	*1/42*	*1/45*	*Post-war*	*Total*
United Kingdom	25	13	13	10	62
United States	14	17	12	13	52
Germany	20	12	5	5	39

Source: Lund (1998, p. 99)

Challenges facing early innovators and entrepreneurs

The conditions that led to the spark that ignited heavier-than-air aircraft development came from the alignment by a number of entrepreneurial individuals of several technologies (balloon, etc.) from related disciplines (e.g. engineering, aviation) and industries, e.g. locomotives, cycling, sailing and motorcars. Table A.1 in the Appendix provides an analysis of the technological background roots of several entrepreneurs prominent at this time and where it has been possible to identify their links to Aero Clubs (which could loosely be described as trade associations). Chapter 3 discusses the technologies coming out of adjacent industries and their role in the development of the emerging aircraft industry in more detail, as well as the general context in which these aviation entrepreneurs operated.

During the inception and fermentation stage of the aviation industry, two key sources of challenges faced early entrepreneurs and their start-up companies – firstly, how to combine the technology to create an imitable aircraft design that delivered reliable and safe aircraft acceptable to the customer base, and secondly, how to keep start-up organisations going until a regular customer base could be established. This is not an unusual combination of problems for entrepreneurs operating in emerging industries. Often it takes time to develop the products that customers are ultimately happy with. Combining technologies in such a way that produces these products can be expensive and time-consuming.

In the case of aviation, major gaps in codified knowledge existed. Those early days saw the entrance of individuals with high levels of technical skill, acquired through undertaking technical college/institutes engineering diplomas, industry apprenticeships or direct technical experience. This led to the situation where, by the late nineteenth and early twentieth centuries, a number of entrepreneurial innovators had demonstrated flight (e.g. the Wright brothers and Whitehead in America and Richard Pearse in New Zealand), although who flew first, and the duration of these flights, are hotly debated as they were only witnessed by small groups of largely uninfluential people. Key technological challenges required a combination of knowledge of the aerodynamics of flight, the development of the means to control and steer flying machines, and the development of light, petrol fueled engines that could provide the necessary power-to-weight ratio required to deliver sustainable flight.

Insights from these early innovators were selectively shared within the aircraft development community. For example, the Wright brothers protected their technological insights initially by keeping their experimentation from the public eye and later through the use of patents. They shared their knowledge with only a small group of other innovators, notably Chanute from France, who later went on to publish and give a talk in Spring 1903 on the Wright's achievements at the French Aero Club in Paris, which caused quite a stir. Initial seed money came from their thriving bicycle business, which provided sufficient money to build and trial their prototypes without involving investors. This was not true of Whitehead, who needed external funding, and who early on shared his ideas with the Wright brothers at a critical stage in their early development work. This has led to an ongoing controversy in the United States around who is the actual father(s) of American flight.

Significant doubt by governments and the military in many countries of the ability of heavier-than-air aircraft to deliver sustained controlled flight meant money from these sources was hard to come by. This is illustrated by the Wright brothers' inability to engage the United States military, despite being arguably the first inventors to attain sustained flight. Conservative military thinkers of the day viewed the aviation community as primitive and dangerous (Jungdahl, 2013). Competing technologies, such as airships, offered superior range, flight duration and payload capacity, which the early aviation innovators were unable to match. In addition, the plane presented a direct challenge to existing military doctrine, organisation and tactics of the day, disrupting traditional service structures by competing with existing technology used by Army and Navy units. It also challenged the traditional roles of Army and Navy units – a problem subsequently resolved by the addition of aircraft units. In addition, fallout from previous failed attempts by governments at using public funds for aviation projects had led many governments to outsource the risk of prototyping and testing to the private sector. Early inventors, therefore, had to acquire money from elsewhere, e.g. from family, from jobs in other industries, from wealthy aviation enthusiast benefactors or aviation/technology displays and competitions. None of this is unusual in emerging industries. Established companies are often too reluctant to finance the development of new technologies because of the uncertainty and risk involved.

The role of the public, media and air clubs in promoting and stimulating aircraft development

The Wright brothers' demonstration of heavier-than-air flight in America in 1903, their subsequent inability to secure a United States Government contract for their plane, and their decision to fly their aircraft in Europe were key in demonstrating to a wider audience the potential of sustained heavier-than-air flight. Pre-1909, few individuals had personally witnessed or experienced heavier-than-air flight, so the capabilities and limitations of the technology were largely unknown. This was not helped by the flimsy appearance of the planes. Wilbur Wright's demonstration of sustained air flight in Hunaudières, France, in 1908 demonstrated the plane's turns and manoeuvres before softly touching down. This shocked many, as sustained flight had not yet been demonstrated in Europe at the time. The Wright flight was followed by two major events in the summer of 1909 that drove the number of confident flyers and their machines from single digits to thousands. The first 1909 event was Blériot's crossing of the English Channel, which demonstrated the possibility of the technology for war. The second 1909 event was the massive air exhibition at La Champagne outside Reims, France. The Grand Aeronautique brought together top aviators (minus the Wright brothers) for several days of aviation competitions and exhibitions, culminating in a major race with a US$10,000 prize. At this exhibition, aviation records in duration and speed were broken several times over. Both 1909 events showcased the capabilities and safety of heavier-than-air flight. Individual aircraft ownership spiked as the European elite turned their attention to aerial competitions and purchasing aircraft copies, e.g. the Glen Curtiss Reims Racer. At the same time, the Wright brothers set up the Wright Company in Germany with local investors and produced models for sale. The Wrights further drove demand by touring Europe, giving their own demonstrations to public and private citizens. From 1908 to 1912 there was rapid acceptance and acceleration of aviation technology, with the public becoming wildly enthusiastic, evidenced by enormous turnouts at air events.

Media moguls, who moved in the social circles of the wealthy, worked to expose and engage the public in the daring exploits of the early inventors. This was at a time when newspapers were changing their dry, factual reporting style to news stories designed to extend their readership to the upwardly moving working and growing middle classes. They published articles promoting and discussing early air shows, putting up, and encouraging others to put up, prizes for the breaking of aviation limits, e.g. flight height, duration, etc. These shows, besides providing innovators with important short-term income boosts, brought inventors and the public together, stimulating information creation and exchange. Early pioneers achieved celebrity status, attracting others into the industry who made aeronautical advances, which drew in more advanced and replicable models.

The champions of aviation promotions were the daily newspapers, e.g. *The Daily Telegraph*, *The Berliner Lokal Anzeiger* and *The Herald*, and the Aero Club magazines (see Flightglobal archive), which covered major aviation events and published regular updates on the progress of inventors and their inventions. Several newspapers produced editorials imploring their national governments to take notice of the advances in aircraft design and capability. Newspapers offered substantial prizes to stimulate aviation record breaking, e.g. the *Daily Mail* offered £10,000 for the first British aviator to fly from London to Manchester. The Aéro-Club de France offered the Grand Prix d'Aviation Deutsch-Archdeacon of 50,000 francs. The *New York Times* awarded US$10,000 for the first flight from New York to Philadelphia and back. In 1910 William Randolph Hearst offered US$50,000 for the first pilot to traverse the United States coast to coast in less than 30 days. For comparison, the United States Army aviation budget in 1912 was only US$125,000.

A series of fictional work centred on aeronautical subjects, particularly the military aspects, also tended to raise public interest. Examples include Jules Verne's *The Clipper of the Clouds* (1886) and *The Master of the World* (1904), George Griffith's *Outlaws of the Air* (1895) and H.G. Wells' *The War in the Air* (1907). A number of these were serialised in newspapers, which gave them further exposure. For example, Wells' work was serialised in *The Pall Mall Magazine* in 1908 and both fascinated and terrified the British public, contributing to a phantom airship scare of 1909 in which a series of zeppelin and aircraft sightings in southern England set off a national wave of panic. Mass media outlets, newspapers, novels and aviation journals therefore contributed to inciting fear and awe at the wonder of flight.

A group of innovative inventors, wealthy enthusiasts, media moguls and concerned citizens raised concerns about the lack of military and government engagement in a number of countries. Working together, they enlisted the mass public in fundraising and letter writing campaigns to pressure national governments to advance the cause of military aviation. Lobbying started in the early 1900s, but by the end of the decade, there was still a reluctance to adopt aircraft in both America (1908) and the United Kingdom (1909). Government and senior military figures were not yet convinced of the practical, offensive utility of aircraft for war, compared with lighter-than-air alternatives. Several early aviation enthusiast campaigns sought to engage lacklustre governments by raising donations for aircraft and equipment for national military services.

Despite the absence of government support, globally, private citizens were enthralled by heavier-than-air flight. In the United States, France, Germany, the United Kingdom and elsewhere, wealthy enthusiasts funded early aircraft development, commissioning and funding early prototypes. The Fédération Aéronautique Internationale (World Air Sports Federation) was founded in 1905 to maintain world records for aeronautical activities. Countries with early Aero Clubs were Belgium (founded 1901), France (founded 1898), Germany (founded 1905), Italy (founded 1904), Switzerland (founded 1900), the United Kingdom (founded 1901) and the United States (founded 1905), profiling where interest in aviation lay, as well as key players and early enthusiasts within the broader community. For example, in the United Kingdom a shift can be seen in those who joined the Aero Clubs from 1908 to 1912 – from early inventors and interested industrialists to aviation enthusiasts and military personnel.

Several clubs, leagues and associations decided early on to work together to maximise their influence. For example, in April 1909, the Aero Club of the United Kingdom, the Aeronautical Society of Great Britain and the Aerial League of the British Empire clarified the complementary roles of each organisation. The Aeronautical Society promoted scientific and technical aspects of aviation, contributing to an expansion of print media outlets focused on aviation developments, particularly aviation related journals, magazines and bulletins like *Flight, Aeronautics, L'Aerophile* and *Flugsport* which tracked aerial achievements, technological progress and upcoming events and made them accessible to a wide range of aviation enthusiasts. The Aero Club trained pilots and studied the art of flying. The Aerial League lobbied the government on aviation issues. Together, this early technological incubation and community engagement was crucial in setting the stage for the explosion in aviation technology.

The involvement of the military and their shaping of the aviation market

By 1910, the first military aircraft was purchased and delivered to the United States Army. Two months later, the French War Ministry purchased three military aircraft of varying designs. The United Kingdom, Germany and Russia all followed shortly after. At the beginning of 1909, there were zero military aircraft worldwide, but by 1910 there were 50, and by 1911 aircraft were being used in military combat for the first time (by the Italians in Libya). By 1914, there were an estimated 1,000 operational military aircraft in service, despite there still being pockets of government and military resistance. As part of this process, fixed-wing aviation was left largely to the private sector inventors and manufacturers. Demand by the military led to a wide range of designs and purposes, allowing organisations to carve out specialist niches in the 1919 to 1939 period. However, funding for the development of competitive commercial airline companies was limited, leading to a slow development in this area.

From an innovation perspective, the development of the aircraft industry could be defined in stages. Based on the Military Factory website, which provides an overview of primarily military aircraft, the following insights can be gleaned.

- Stage 1 (1900 to 1909 – inception and experimentation) – small number of companies prototyping and demonstrating a small number of designs and planes. Military not engaged fully in the industry, media and public arena. Formation of Aero Clubs. The United Kingdom, the United States and France were key players.
- Stage 2 (1910 to 1919 – immediate pre-war years and World War I) – dramatic increase in the number of companies entering the market and designs being offered. Over 1,800 per cent explosion in the number of companies in market and over 5,000 per cent explosion in the number of aircraft available. There was a 1,300 per cent explosion in the number of designs and prototyping. Dominant plane areas were fighter, bomber, multi-modal and reconnaissance

planes. Lots of small organisations existed. Germany dominated production, followed by the United Kingdom, France, the United States and others. Employment grew to very substantial levels and production was conducted on a large scale (see Figures 2.1 and 2.2).

- Stage 3 (1920 to 1929 – interwar years) – consolidation of designs and dramatic drop in the number of organisations producing planes; also a drop in the amount of prototyping. Introduction of transport planes. Fifty per cent drop in the number of producers and 73 per cent drop in aircraft designs

Figure 2.1 Avro female workers at Brownsfield

Figure 2.2 Armstrong Whitworth Works at Whitley

produced by these organisations. During this period the United States, the United Kingdom, the Soviet Union and France dominated production. The Czechs entered the market.

- Stage 4 (1930 to 1939 – pre-World War II years) – dramatic increase in the number of producers and aircraft designs. Limited experimentation and prototyping, build-up of planes of war, e.g. fighter, bomber, multi-modal and reconnaissance planes. Slow development of transport planes. Large organisations dominate. There was an increase of 240 per cent in the number of producers and 360 per cent increase in aircraft designs being produced. Aircraft production was dominated by the United Kingdom, the United States, Germany and the Soviet Union, with France, Italy and Japan following.
- Stage 5 (1940 to 1949 – World War II and immediate post-war years) – explosion of the amount of experimentation and prototyping primarily in the military plane space. A return to large-scale production of planes of war. Transport plane development took place, similar to what occurred in Stage 4. Large organisations dominate. There was a 136 per cent increase in the number of producers and 260 per cent increase in aircraft designs being produced. Aircraft production was dominated by the United States, Germany, Japan, the United Kingdom and the Soviet Union with France and Italy following.
- Stage 6 (1950 to 1959 – post-World War II) – halving of the number of producers in the marketplace and number of aircraft designs being produced. Focus on fighters remains. Development of transport plane industry. Large organisations dominate. A 53 per cent drop in the number of producers and 70 per cent drop in aircraft designs being produced. This period was dominated by the United States, the United Kingdom, Russia and Canada.
- Stage 7 (post-1960) – focus on fighters. Development of transport plane industry. Large organisations dominate. Continued consolidation and drop in the number of produces and aircraft designs being produced.

The realisation of war drove the growth of the market in Stages 2, 4 and 5. Post-war, consolidation occurred, leading to a rationalisation of producers and designs (Stages 3 and 6). In Stages 1 to 2, the search for a dominant design drove exponential growth in designs and companies. Once dominant players and designs emerged, the range and number of designs fell, although emerging new technologies such as retractable undercarriages and jet engines did bring design modifications. Fighter plane production remained steady throughout all stages, with commercial and transport planes slowly emerging as a dominant industry in the latter stages.

A deeper analysis of these stages shows the drivers for changes in the market. In 1910 to 1919 (Stage 2), aircraft, initially unarmed, were used extensively in reconnaissance until personal weapons were added, leading to the emergence of the fighter plane, which helped to overcome the stalemate of trench warfare via air superiority over the enemy. Once the military realised the potential of flight for warcraft, they invested in heavier-than-air flight research and development. This brought about a rapid change in market shares and leading companies and products, particularly during the 1930s, with the build-up to World War II (Stage 4). War accelerated the rapid diffusion of new technologies to aircraft design, launch and commercial producing, increasing costs and leading to fewer but larger producers.

In the post-World War I (Stage 3) period, the dramatic drop in the need for military aircraft led to a partial collapse of the industry as aviation companies struggled to find new markets. Some governments (e.g. the United States) provided support to a few organisations to move into the commercial aircraft market, but this was not universal across countries. In the United Kingdom, diplomacy with Germany meant no investment in the air industry until World War II was imminent. France soon fell behind the United States and Germany due to an underinvestment in public research, engineering schools and design. In Germany the Treaty of Versailles (1919) imposed a ban on research and the building of military aircraft, resulting in aircraft companies converting to civil aviation.

In Stage 4 (pre-World War II), the United States maintained its dominance of the production of commercial aircraft. In contrast, in 1939, the British military aviation industrial complex was more efficient than either that of the United States or Germany. The British national industry possessed the greatest production experience, largest order book and strongest research and development side (Lund, 1998). The United Kingdom's island status, its imperial liabilities and its proximity to its German foe dictated a particular strategic approach to the air war that resulted in the Royal Air Force generously supporting industry and research. This favoured a wide range of designs, necessary for the myriad of purposes demanded of British planes in wartime, e.g. strategic photography, weather reconnaissance, air search and rescue, electronics intelligence and electronic warfare and the production of quality aircraft over quantity. This allowed the United Kingdom to develop a specialist niche market in military work from 1919 to 1939.

In Germany, Hitler's appointment as chancellor of Germany in 1933 meant that the German Government began to pour government support into research, engineering schools, design and private companies. This led to Germany emerging as a leading developer and producer of planes. In the Soviet Union at the same time, an aviation manufacturing complex was built up of considerable size, while in Japan and Italy, similar military focused industries were expanded.

In Stage 5, the United States became a major market for British aviation technology, particularly Rolls-Royce Merlin engines, aviation equipment, radio, armaments, cannons, gunsights and other components. Aero engine manufacturers in the United Kingdom (Metropolitan Vickers, de Havilland and Rolls-Royce) led the development of jet engines, with the United States and Germany also involved in the area. Germany found that sharp-wing sweepback improved aircraft performance at high subsonic speeds, but were unable to capitalise on this as their jet engines were not yet powerful enough to realise the potential (Lund, 1998). In this area, the United Kingdom dominated (see Table 2.1).

After World War II (Stage 6), Germany's aircraft industry was destroyed and the United Kingdom and France were consumed with reconstruction and debt, leaving the United States to emerge as the aviation industry leader and to benefit from complementary technologies and industries.

In some countries, a ban was placed on foreign companies entering the local market, stifling the early exchange of innovations between the United States, France, Germany and United Kingdom. In the early stages of aviation in the United Kingdom, it was relatively easy for local enterprises to get up and going as the policy of air and naval authorities was to experiment widely with multiple entrepreneurs on aircraft designs. World War I transformed the industry from a small-scale cottage industry into a large-scale, complex one.

Through all of these various stages, technological developments drove the growth of the industry, and entrepreneurs played an important role in initiating much of these developments. Aviation in the period up until World War II was emerging and unstable – circumstances that created a variety of opportunities for those prepared to take them on. Public support and government funding were critical in converting opportunities to reality.

2.5 Conclusion

Entrepreneurs tend to thrive most where there are emerging industries that provide them with the opportunities to strike out on their own. Understanding the nature and patterns that arise amongst these entrepreneurs in their emerging industries can help to provide a better understanding of the entrepreneurial process. It also helps to provide insight into the factors that are likely to contribute to the transition of organisations from start-ups to dominant players in those industries as they evolve.

In the specific case of aviation, the period from 1900 to 1910 marked the beginning of the public's awareness of heavier-than-air flight and the promise it offered, but the military initially did not get on board until just before the start of World War I. Early investment and stimulation for the emerging aviation industry came mainly from the general public. Subsequent military contributions before both world wars delivered two important fermentation periods, characterised by an explosion of start-up companies experimenting with and delivering multiple different designs. World War I delivered the first round of early start-ups. World War II saw more dominant players thrive and non-entrepreneurial but well-run companies from other industry sectors move into the aviation space.

The post-war years and the withdrawal of military money helped consolidate designs and companies to a few key players who subsequently went on to dominate the further development of the industry. The ebbs and flows of military money both enabled and constrained entrepreneurial activity. In the early stages, entrepreneurs relied heavily on air shows and family for start-up funding. Military money allowed a number of these organisations to grow but also shaped their efforts. As technological complexity increased, the need was greater for sophisticated business management skills, which the more successful entrepreneurial companies had. Those unwilling or unable to ramp up their business management capability exited the industry, resulting in a legitimate and credible design and manufacturing base which dominated further development. Despite all of these changes, however, a number of key people in the early days of the industry took advantage of their flying achievements to found companies and then lead them for a number of subsequent years. They did so without uniform success, but their efforts contributed to the creation of an industry and a range of companies that is still with us today.

References

Aldrich, H. (1999). *Organisations Evolving*. London: Sage.

Aldrich, H. E. and Fiol, C. M. (1994). Fools rush in? The institutional context of industry creation. *Academy of Management Review*, 19(4), pp. 645–670.

Aldrich, H. E. and Martinez, M. (2003). Entrepreneurship as social construction: A multi-level evolutionary approach. In *Handbook of Entrepreneurship Research*. Boston, MA: Springer, pp. 359–399.

Anderson, P. and Tushman, M. L. (1990). Technological discontinuities and dominant designs: A cyclical model of technological change. *Administrative Science Quarterly*, pp. 604–633.

Flight Global Archive. Available at: www.flightglobal.com/pdfarchive/view/1909/1909%20-%200255.html

Forbes, D. P. and Kirsch, D. A. (2011). The study of emerging industries: Recognizing and responding to some central problems. *Journal of Business Venturing*, 26(5), pp. 589–602.

Funk, J. L. and Methe, D. T. (2001). Market and committee-based mechanisms in the creation and diffusion of global industry standards: The case of mobile communication. *Research Policy*, 30(4), pp. 589–610.

Garud, R. (1994). Cooperative and competitive behaviors during the process of creative destruction. *Research Policy*, 23(4), pp. 385–394.

Goldberg, A. (1955). Equipment and services. In: W. F. Craven and J. L. Cate, eds., *The Army Air Forces in World War II*, Vol. 6. Chicago, IL: University of Chicago Press, pp. 350-352.

Jungdahl, A. (2013). Public influence on the proliferation of military aviation 1907–1912. *Air Power History*, 60(1), pp. 28–39.

Lund, E. (1998). The industrial history of strategy: Reevaluating the wartime record of the British aviation industry in comparative perspective, 1919–1945. *The Journal of Military History*, 62(1), p. 75.

The Military Factory – Aircraft by Decade. Available at: www.militaryfactory.com/aircraft/aircraft-by-decade.asp

Powell, W. W. (1990). Neither market nor hierarchy: Network forms of organisation. In: B. M. Staw and L. L. Cummings, eds., *Research in Organisational Behavior*, Vol. 12. Greenwich, CT: JAI Press, pp. 295–336.

Rosenkopf, L. and Tushman, M. L. (1994). Technology and organization. *Evolutionary Dynamics of Organizations*, p. 403.

Sarasvathy, S. D. (2009). *Effectuation: Elements of Entrepreneurial Expertise*. London: Edward Elgar Publishing.

Van de Ven, A. H. (1991). *A Systems Framework for Studying the Process of Entrepreneurship*. Paper presented at a conference on Theories of Entrepreneurship. Champaign-Urbana, IL: University of Illinois.

3 Overview of the global aviation industry 1903 to 1945

3.1 Introduction

Looking at the behaviour of entrepreneurs requires a strong understanding of the nature of the industries and markets in which they operate. In the case of high-tech, dynamic industries, where strong improvements in technology drive development, it seems obvious that these markets provide opportunities for new entrants. Technological change disrupts old business dominances and opens up opportunities to newer entrants to prosper. It also helps to create demand for new products, which can help new entrants.

This, however, is not the entire story. Technological change can alter the nature of market and industry structure, and the demand for economies of scale and scope in production. This can create barriers to entry. The changes that occur can end up meaning that the knowledge acquisition and capital expenditure required for entry is prohibitively high. Early entrants to an industry may develop an advantage that enables them to dominate the industry long after the initial phase of technological innovation has passed. Understanding how these changes occur and why is necessary if a study of the nature of the behaviour of entrepreneurs is to be undertaken.

In the case of the aviation industry technological change was rapid after 1903. The rate of this change did at times accelerate, but up until 1945, there were no sub-periods of technological stability. In this period, there was significant development in the nature and structure of airframes and aero engines but also in the development of a whole range of associated products such as instruments, undercarriages, radio equipment, weapons, etc. This meant that a variety of new companies were created, not only in the manufacture of aero engines and airframes but also in a range of other associated products.

The purpose of this chapter, therefore, is to provide an account of the general changes that occurred in the industry, both in terms of technological change and in terms of construction organisation and demand. The chapter is divided into historical periods: pre-World War I, World War I and the interwar period. Finally, a section is provided with information on the development of commercial aviation.

3.2 The origins of the aircraft manufacturing industry (1903–1914)

The manufacture of aircraft began in the United States with the work of aviation pioneers such as Wilbur and Orville Wright and Glenn H. Curtiss. Initially, the United States led world innovation in the development of heavier-than-air aircraft at the start of the twentieth century; however, they were not to retain this position for long. During this period, the industry in the United States remained a highly competitive one but tiny in size and in nature a specialised cottage industry that designed and built in small workshops a small number of handcrafted aircraft for wealthy enthusiasts (Van der Meulen, 1991). The handicraft nature can be discerned from looking at Figure 3.1, which shows a pre-World War I aircraft being built. Aircraft were, however, difficult to sell as they were relatively expensive, had a limited payload and were dangerous to fly.

Between 1903 and 1914, aviation pioneers in both the United States and Europe turned out to be very innovative and managed to create spectacular advances in performance capabilities. For instance, in December 1903, the Wright brothers made their first three flights, which were all under one minute's duration and travelled a distance of less than 1,000 feet.[1] By 1914 (about ten years later), pioneer aviators were flying aircraft at ranges of over 600 miles, with top speeds of nearly 130 miles per hour and were reaching altitudes of over 20,000 feet. This meant that although aircraft had few uses initially, their usefulness quickly was extended.

Despite the efforts of the Wright brothers to retain knowledge of flight to themselves – they largely did not publicise their feats before 1908 – flight soon emerged in other countries. The first flight in Europe was undertaken in France in 1906 by Alberto Santos-Dumont, a Brazilian émigré living in Paris, and in the United

Figure 3.1 Nieuport production at Cheltenham, England

Kingdom, the first powered flight was made at Farnborough in October 1908 by J.T.C. Brabazon.

The European and American aviation industries, therefore, began to develop within a few years of each other. In each case, the first formal steps to establish companies dedicated to the manufacture of aircraft began in the years just prior to World War I. The technical lead, which the United States had through the pioneering efforts of the Wright brothers, very quickly evaporated and the focus of aviation development shifted to Europe. France initially led the way in heavier-than-air flight, and by the outbreak of World War I in 1914, French flyers held all of the world records in aviation. In contrast, Germany had a lead in the development of airships, but like France, its commitment to aviation overall was stronger than in the United States and the United Kingdom. The military market, even before 1914, was of great importance to the early aviation innovators and entrepreneurs, although the British and American governments were slow in this field, compared to that of France, Germany and even Russia. Table 3.1 provides data on the numbers of aircraft possessed by a range of countries in 1914. As can be seen, these figures are quite small, and it is notable that even the militaries of countries like Austria and Italy had more aircraft than that of the United States. In the case of the United States, the numbers produced at this time were tiny – 29 aircraft in 1912, 29 again in 1913 and 34 in 1914 (Bilstein, 2001, p. 225).

Despite these very small numbers, important companies were formed in this period. The Wright brothers formed their company to produce aircraft in 1909 and Curtiss formed his first company in 1910 (the two later merged in 1929, and the merged entity – Curtiss-Wright – still exists today). In the British case, before 1914, some British firms were agents assembling French machines, and those firms that did design their own airframes were wholly dependent on foreign (usually French) engines. In the United Kingdom, the Short brothers and Handley Page formed aviation manufacturing companies in 1908, followed by Blackburn and A.V. Roe in 1910, Sopwith and British and Colonial (later Bristol) in 1911. Despite the British industry's reliance on French technology, the performance of British aircraft improved steadily. Before the war, many firms in the United

Table 3.1 Military aircraft inventories, 1914, 1918, 1927/28, 1939

	Front line 1914	*Production Word War I*	*Front line 1918*	*1927/28*	*1939*
France	138	67,987	4,500	3,500	1,735
United Kingdom	113	58,144	3,300	744	1,750
Germany	232	48,537	2,390	0	3,960
Italy	150	20,000	1,200	N/A	N/A
United States	55	15,000	740	1,857	2,400
Austria	86	5,431	–	–	–
Russia	244	4,700	579	914	8,139

Note: Aircraft are front line aircraft in each case, not training types.

Source: Angelucci (2001)

Kingdom were very small, making no more than one of two aircraft, and the more successful ones were not much bigger. The output of the British and Colonial Aeroplane Company, for instance, was approximately 200 machines between 1910 to 1914, with A.V. Roe building approximately 70 and Handley Page fewer than a dozen (Fearon, 1969, p. 487).

The pioneers whose companies survived in the United Kingdom after World War I, such as the Short brothers, A.V. Roe, Sopwith, de Havilland, Fairey, Blackburn and Handley Page, all had in common a training in engineering and their firms were financed privately. This finance came from their families, private individuals, prize money, the operation of flying schools and to a limited degree the sales of aircraft. Support from established engineering firms was very limited but there were some enthusiasts in heavy engineering such as those at Armstrong Whitworth, Vickers and George White's British and Colonial.

The British military did purchase an increasing number of aircraft between 1912 and 1914; however, their purchases were modest in number. In the United Kingdom, as in many countries, military interests did to a certain degree drive the birth and development of the early industry (Owen, 1999, pp. 296–297). The first aircraft flight in the United Kingdom took place at the Farnborough military base in 1908, and in 1912, the first orders for military aircraft took place; the same year of the creation of the Royal Flying Corps. The Royal Naval Air Service was formed in 1914 when the Navy began buying aircraft.

In the United Kingdom, technological developments were stimulated by government involvement through the high-quality aeronautical research undertaken at the Royal Aircraft Establishment and National Physical Laboratory and at a few universities, mainly Manchester and Cambridge. As previously mentioned, the founders of the early aviation companies were often well-trained engineers, and in a number of cases, they could be described as entrepreneurial in nature. These individuals were in love with flight and the promises the technology offered for the future. These early aviation pioneers, or entrepreneurial engineers (such as Frederick Handley Page, Thomas Sopwith, Robert Blackburn and A.V. Roe), were to play an important role in the development of the British aviation industry.

At the same time that these events were happening in the United Kingdom, similar types of individuals were achieving comparable outcomes in other countries. In the United States in the years after the Wright brothers' first flight in 1903, around a dozen companies were founded. As elsewhere, these were small firms producing mainly experimental types developed by the individual. Amongst the new entrants besides the Wrights and Curtiss (mentioned earlier) were ones founded by pioneer aviators, such as William Boeing (1916), Allan and Martin Lockheed (1912) and Glenn Martin (1912). The situation was similar in France where aviation pioneers such as Henri Farman (1908), Louis Blériot (1905), Edouard de Nieuport (1908) and Gabriel and Charles Voisin (1906) all formed aviation manufacturing concerns. Anthony Fokker (1912) was a similar Dutch/German example, along with August Euler (1908) and Gustav Otto (1911) in Germany, Igor Sikorsky in Russia and Umberto Savoia (1915) and Giulio Macchi (1913) in Italy. Like the British pioneers, most of the American, French and

German pioneers had engineering experience, especially in the infant motorcar industry and locomotive industry, and a technical education.

Public support for the new industry was keen and around this time the Fédération Aéronautique Internationale (World Air Sports Federation) was founded (1905) to maintain world records for aeronautical activities. Founding Aero Club members were from France, Germany, Belgium, Italy, Switzerland, the United States and the United Kingdom. The Aero Clubs of the United Kingdom, the United States and France were particularly important to the development of the fledgling aircraft industry as they offered major air shows and competitions with prize money drawing in a wide range of national and international competitors. These air shows drew together aviation enthusiasts from all walks of life, providing an important network enabler and opening up potential pathways and resources to the early aviation entrepreneurs. A combination of aero clubs, sponsored races and exhibitions in America and Europe played an important role in bringing together engineers, aviators, newspaper owners (who often provided substantial rewards for winning aircraft race competitions), industrialist aviation advocates and government representatives. These international events provided networking and technical sharing opportunities.

Out of this melting pot of technological change emerged a number of engineers/entrepreneurs and new, innovative companies. The rapid change in the technology associated with the industry made it particularly attractive to entrepreneurial types of individuals and to young, technically receptive engineers who proved to be fascinated, first with the possibilities of flight and its potential but then later with the challenge of solving the complex technical problems that arose in the industry.

> When an industry is growing it needs the stimulus of alert, flexible, young minds; it needs vision, courage and drive.
>
> (Handley Page, 1944, p. 409)

Interesting and challenging work for technically capable individuals were abundant, which attracted individuals intensely interested in the opportunities available in this emerging, tiny, well-connected, dynamic industry ecosystem. Those early days saw entrance into the industry of individuals who had high levels of technical knowledge and skill, generally acquired from technical college engineering diplomas and industry apprenticeships (often from both) as well as experience in engineering. Initially, these innovators created small firms, often funded from competition prizes in aircraft and related technology areas. These firms quickly were transformed into substantial enterprises fueled by the rapid expansion of military demand during World War I. Despite these early aviation entrepreneurs having a solid engineering background (often electrical or in the motor trades), they generally had little business training, or even experience. Business skills were, therefore, often learnt on the job, mainly through trial and error, rather than from any sort of formal training. Fortunately, the amounts of money needed to fund these early concerns were not prohibitive. In the early days of the industry, the need to create substantial economies of scale in manufacturing aircraft was not great, as they were generally made from wood and could be constructed by relatively small firms, meaning the start-up costs were fairly modest.

Besides the military, the other main source of revenue for these companies before 1914 was private aviation enthusiasts wanting to fly and have their own planes. This helped created a limited demand for aircraft but also created another limited stream of income for the aviation pioneers. Many of the early pioneering aviation manufacturers also owned and operated flying schools. Any possible further development of demand, however, was limited as aircraft were risky to fly and expensive to purchase and maintain. This situation was to entirely change during World War I as national industries geared up to produce as many aircraft as physically possible.

3.3 World War I (1914–1918)

The outbreak of war in 1914 led to a burst of growth in the industry. The progression of the war eventually led to governments placing an increasing emphasis on air power. As demand by the military picked up during World War I, the industry began to be transformed from that of a small-scale cottage industry into a huge and complex industry. The number of firms in the United Kingdom and France engaged in the industry eventually ran into hundreds; however, after the war, most of those who had come from associated industries returned to their pre-war work, leaving the aviation industry generally to the pre-war aviation manufacturers. During the war itself, aircraft manufacturing expansion was stretched to almost its physical limits in the United Kingdom, France and Germany, which meant that after the war there were stockpiled large numbers of surplus aircraft. Many of these were eventually scrapped.

The rapid expansion of the industry during World War I also created tremendous opportunities for the pioneer flyers. A number of relatively young men were thrust into the position of managing industrial enterprises of considerable size, based on the fact that only they knew much about the design and manufacture of aircraft. In the United Kingdom, for instance, by 1918, the young 30-year-old flyer Thomas Sopwith had become the managing director of a company (Sopwith Aviation) employing over 5,000 people. These sorts of responsibilities encouraged a number of these pioneer aviators to struggle on in the industry once the wartime boom was over.

To gain an appreciation of the scale of the wartime operations Table 3.1 shows production figures for aircraft in World War I, with the French industry producing the most, followed by the United Kingdom and Germany. In the United Kingdom the leading industry manufactures grew into large enterprises, with the biggest, the Aircraft Manufacturing Company (Airco), employing 7,000 people by 1918 (Owen, 1999, p. 297). At the same time, the British RAF had 27,000 officers, 260,000 other ranks and over 3,000 aircraft (Dobson, 1991, p. 9). The British industry at the end of the war employed 112,000 people, producing on average 1,250 aircraft per month (Hayward, 1989, p. 11). In France at the same time, production in 1918 was even greater at about 2,000 per month (Chadeau, 1986, p. 435).

As the United States did not enter the war until 1917, production did not increase in that country substantially until that year. Once the United States did enter the war, production increased substantially, and between April 1917 and November 1918, 13,894 aircraft and 41,953 engines were delivered (Rae, 1968, p. 2). During World War I, few non-military aircraft were built outside of the United States; instead, the aviation industry focused on meeting the demands of the war. The war did, however, drive technological advancements, which added integrity and robustness to aircraft design generally.

In World War I, the United States did not design and manufacture any aircraft of its own for use at the front. Instead, most wartime manufacturing focused on the production of training aircraft – the British-designed DH-4 fighter – and aircraft engines, with the motorcar companies dominating. In the United States, peak production was reached late in 1918, when the industry employed 175,000 people, at a rate of 12,000 aircraft per year. The largest military contracts for producing the DH-4 went to Dayton-Wright in Ohio, Glenn L. Curtiss, Fisher Body and Standard Aero. In Europe itself, most United States airmen used French-made aircraft in combat.

This expansion in American production meant that the crippling patents dispute surrounding the use of the Wright brothers' original designs that impeded the growth of the industry had to be settled. This was accomplished through the establishment of the Manufacturers Aircraft Association, whose members agreed to a cross-licencing arrangement that allowed manufacturers to use the technology that others had patented for a fee. The largest of these fees were paid to the Curtiss and Wright companies.

3.4 The interwar years (1919–1939)

Immediately after the end of World War I (in the early 1920s), the aviation industry in all countries went through a severe and painful contraction. Military contracts from government dried up quickly, and for a period, governments did not make many ongoing purchases at all. This was because there was at this time a surplus of fuselages and engines left over from the war and so governments made use of these to meet their greatly reduced demands. To commercially survive, some aircraft manufacturers tried to diversify their production into a range of different fields including such things as the manufacture of furniture, motorcars, bikes and even such things as milk churns. In a number of cases, some of the very large concerns were simply unable to adjust and survive and ultimately disappeared altogether. For example, in the United Kingdom the company Airco was initially sold to the parent company of a manufacturer of motorcars before finally being liquidated and its assets sold off. Its aviation manufacturing assets were sold to Airco's chief engineer, Geoffrey de Havilland, who used them to establish his own aviation manufacturing enterprise (Fearon, 1969, pp. 476–479).

With the severe reduction in demand for aircraft, the British industry became a relatively small one, dominated by a group of companies run by very dedicated pre-war producers. In addition, there were a few new enthusiasts whose interest had been aroused during the war. This trend also took place in the United States and France, where many aircraft manufacturing firms tried to diversify their operations in order to maintain their organisations, some making motorcar parts while others ventured into commercial aviation.

After the war's end, there was a large number of cheap surplus aircraft (as well as trained pilots) available, and this helped to bring about an increase in private flying, joy riding and taxi work. Wartime aircraft, however, proved to be unsuitable for economical commercial flying and it became apparent that for successful private flying, specially designed light aircraft were needed. In the United Kingdom, this demand was largely met by the development of the de Havilland Moth in 1924, and so successful was this aircraft and other Moth types that by the mid-1930s they accounted for half of the civil aircraft registered in the United Kingdom (Fearon, 1969, p. 477). Similar types of aircraft were produced as well in France and the United States.

After the post-war contraction there remained a group of small and financially strapped but enthusiastic producers in each country. In the United Kingdom, the military continued to sponsor, through tiny orders, a small number of producers in preparation for new hostilities. Table 3.2 provides data on the number of aircraft in the United Kingdom and France in 1922 and 1925. As can be seen from the figures, aircraft numbers fell considerably from their wartime peaks. Production levels and employment levels also shrunk considerably (see Table 3.5). By the mid-1920s the post-war shake out of the industry was over and demand and production rose in most countries through to 1929. Tables 3.4 and 3.5 provide employment and production figures in France, the United Kingdom and the United States. As can be seen from the figures employment in France, the United States and the United Kingdom were all substantially higher in 1929 than in 1924, a reflection of the recovery of the industry.

Although the average number of orders for military aircraft in each country were initially small, sales increased from 1923 on. By the time of the British Census in 1930, there were still 38 companies in the airframe manufacturing industry in that country, although 16 of them did account for approximately 90 per cent of all production (Fearon, 1974, p. 237). This number did not, however, remain stable. During the 1930s, further technological changes created pressure for these companies to merge into larger establishments.

Table 3.2 Military and civil aircraft in the United Kingdom and France, 1922, 1925

	United Kingdom	*France*
Military (December 1922)		
Home	137	946
Abroad	234	232
Total	371	1,178
Estimated total (1925)	575	2,180
Civil	167	600

Source: Higham (2012)

Table 3.3 Military aircraft in the United States, 1926, 1927

	1926	*1927*
Aeroplanes	1,125	1,857
Seaplanes	61	105
Total	1,185	1,962

Source: United States, Bureau of Census (1927)

Table 3.4 Estimates of employment in aircraft industries, '000, 1918–1939

	Germany	*United Kingdom*	*France*	*United States*
1918		112	170	20
1924	–	12	13	–
1929/30	–	21	20	19
1934	46	24	22	10
1936	188	75	32	–
1938	293	128	47	–
1939		140	110	64

Source: Edgerton (1991, p. 26), Hayward (1989, pp. 11–14) and Rae (1968, p. 102)

Table 3.5 Selected statistics on the aircraft production industry, 1920 to 1939

	United Kingdom		*United States*		*France*	
	Employment no	*Production no*	*Employment no*	*Production*	*Employment*	*Production*
1920	na	na	na	328	5,201	976
1921	na	na	1,952	437	3,711	1,143
1922	na	na	na	263	7,476	1,135
1923	2,500	na	3,496	743	10,647	977
1924	11,735	503	na	377	13,461	1,359
1925	na	na	3,200*	789	15,553	1,478
1926	na	na	na	1,186	15,916	1,650
1927	na	na	5,486	1,995	11,394	1,536
1928	na	na	na	4,346	15,074	1,532
1929	na	na	18,620	6,193	16,098	1,440
1930	21,322	1,456	na	3,437	20,374	1,570
1931	na	na	na	2,800	22,661	1,400
1932	na	na	na	1,396	19,945	1,355
1933	na	633	9,626	1,324	16,922	na
1934	23,774	1,108	na	1,615	20,821	na
1935	35,032	1,807	14,931	1,710	32,643	1,052
1936	75,000	2,218	na	3,010	38,495	1,028
1937	na	2,827	na	3,773	38,495	1,092
1938	128,000	2,800	na	3,623	58,265	1,595
1939	140,000	7,940	63,994	5,856	82,289	3,338

* The number of aircraft exported and ordered by the Air Ministry were higher than the number produced in 1924 because of the wartime stocks still not having been exhausted.

Source: Hayward (1989, p. 14), Edgerton (1991, p. 26), Modley and Cawley (1953), Truelle (1969, pp. 89–102), Chadeau (1986, p. 435)

In the United States, a similar process took place at this time. After the war, employment fell to only 4,000 in 1921, with many companies going bankrupt or being reorganised. This reorganisation saw a number of professional managers replacing some of the early aviation pioneers, leading to improved company profitability. Aviation development was assisted by the United States Government through the passage of the *Air Commerce Act 1926* and creation of the Federal Bureau of Aeronautics. The passage of the *Air Mail Act 1925* spurred the development of new aircraft both for the carrying of mail and eventually for transporting passengers. In addition, the American Army and Navy slowly began to demand more aircraft (see Table 3.3). These changes helped to increases sales and between 1926 and 1928 industry output grew from 1,186 to 4,346 planes. In the United States in the late 1920s and early 1930s, new companies like Fairchild, Grumman, Seversky, the merged Curtiss-Wright firm and North American Aviation were all established. Douglas Aircraft went public in 1928 as well and expanded its facilities, Lockheed re-entered the field around this time and opened a new plant, and Consolidated (which went public in 1929) and John Northrop in 1932 established the Northrop Corporation. Several large holding companies were formed in these years, and although legislation was passed in 1934, which compelled aircraft manufacturers to separate their operations from airlines, a number of those holding companies were to form the basis of the manufacturing companies themselves. In the United States, commercial airline orders grew throughout the 1930s and by 1939 employment in the industry had risen to 36,000.

In the period of the 1920s and 1930s military establishments were aware of the potential of the technology and once convinced of its value drove the early development of the early aircraft manufacturing industry. The demand for aircraft in the United Kingdom, for instance, was maintained by the Royal Air Force in a protectionist environment funded by public budgets which created what could arguably be called the strongest aircraft industry in Europe. Barriers to entry into the British market for non-British competitors were high, despite the exchange of innovations between the United States, France and Germany in the early stages of aircraft development. This meant that within the United Kingdom it was relatively easy for local enterprises to get up and going due to the air and naval policy to experiment widely with multiple entrepreneurs on aircraft design.

The British Air Ministry's decisions to a) spend a large proportion of its budget (about a fifth on prototypes) and b) not to design fighters for more than two years at a time allowed for a greater responsiveness to technology innovations compared with the United States and Germany. This approach favoured the effectuation 'do and review' product development approach of engineers. Despite the low demand in the United Kingdom companies did begin to specialise; Gloster, Hawker and Bristol on fighters, Blackburn, Short and Supermarine on flying boats, Fairey on light bombers and de Havilland on commercial aircraft. A similar level of specialisation took place in the United States and France, with Boeing and Curtiss specialising in fighters and Martin on bombers in the United States and Farman on bombers in France.

Overall demand, however, was still quite low. In 1930, the United Kingdom only had 780 frontline machines compared to the United States' 1,310 (*Daily Telegraph*, 17 April 1930; Dobson, 1991, p. 20). At this time, technological developments were substantial and included the introduction of metal construction, allowing the manufacture of stronger aircraft and the achievement of greater speeds. This change meant that the firms were required to make greater technical efforts than had been the case with wooden construction. The design and production complexity of aircraft was increased, which in turn increased the requirements for more teamwork and the employment of people with a broader skill range. Aircraft design and manufacturing, therefore, began to change from a general reliance on intuitive skills to the introduction of more systematic and scientific approaches to developments. Other major innovations that occurred at this time included the replacement of biplanes with monoplanes, the use of retractable undercarriages and the manufacture of lighter, more powerful engines. Altogether these technological changes helped to create greater economies of scale of manufacturing in the industry. This, in turn, led to a cost of entry to the industry which put pressure on firms to begin to merge. Growth in the industry meant that by 1935, most firms had grown to a reasonable size, with the 11 largest firms in the United Kingdom, for instance, all employing between 1,000 and 2,000 workers (see Table 3.6). Development of the industry accelerated in the late 1930s, driven by the demands of rearmament, and by 1939 in the United Kingdom, there were 130,000 people working in the manufacture of airframes along with an additional 80,000 making aero engines.

Driven by the demands of rearmament and World War II, employee numbers in the United Kingdom rose to 750,000 by 1940. At this time production increased in all countries (except in occupied France; see Table 3.7). World War II also saw the construction of the largest aircraft manufacturing facilities in the United States (see Figure 3.2). When World War II ended, there was another post-war contraction of the industry; however, it was not to be as severe as the one after World

Table 3.6 Labour employed by British airframe manufacturers (April 1935)

	No.
De Havilland Aircraft Co. Ltd.	1,590
Handley Page Ltd.	1,160
Saunders-Roe Ltd.	654
Gloster Aircraft Co. Ltd.	1,200
Hawker Aircraft Ltd.	1,756
Sir W. G. Armstrong Whitworth Aircraft Company	1,092
Others	11,168
Total	18,620

Source: Fearon (1978, p. 81)

Table 3.7 Production of military aircraft, 1939 to 1945

Country	*1939*	*1940*	*1941*	*1942*	*1943*	*1944*	*1945*	*Total*
USA	2,141	6,068	18,466	46,907	84,853	96,270	45,852	300,557
France	3,263	2,113	1,001	1,121	1,444	970	900	10,812
Germany	8,295	10,862	12,401	15,409	24,807	40,593	7,540	119,907
USSR	10,382	10,565	15,737	25,436	34,900	40,300	20,900	158,220
UK	7,940	15,049	20,094	23,672	26,263	26,461	12,070	131,549
Japan	4,467	4,768	5,088	8,861	16,693	28,180	8,263	76,320
Total	33,225	47,312	71,786	120,285	187,516	231,804	94,625	786,550

Source: Bishop (2002); Chadeau (1986)

Figure 3.2 Production of B-24 bombers and C-87 transports

War I because of the development of the commercial passenger industry and the continued Cold War demand for military aircraft.

Besides the growth of demand for military aircraft from the mid-1920s, other developments also occurred. Growth in the amount of private flying made an impact on the industry in the late 1920s when flying as a sport grew in popularity and a number of the smaller firms focused on manufacturing light aircraft. The Great Depression in the early 1930s, however, was particularly destructive to this part of the industry.

The ups and downs in demand in the industry at this time made it a very challenging environment in which to survive. The severe shakeout of the industry immediately after World War I meant that often it was only the pioneering spirit and enthusiasm for aviation amongst the entrepreneurs which kept them in the industry given the poor financial returns they received. One of the United Kingdom pioneer aviators/entrepreneurs when later writing about the period stated, "During this early period, it was a case of money going out and very little coming in. Certainly, we made no profits" (Roe, 1939).

Overall, the dilemma faced by those early firms was how to combine technical innovative flair with the ability to create, finance and manage firms of increasing scale. In the case of the United Kingdom, its island status, its imperial liabilities and its proximity to European rivals dictated a particular approach that resulted in the RAF generously supporting industry and research. This strategy favoured a wide range of designs necessary for the myriad of purposes demanded of British planes in wartime (e.g. strategic, photographic, weather reconnaissance, air search and rescue, electronics intelligence and electronic warfare) and the production of quality aircraft over quantity. The basis for a flourishing export trade in military aviation technology emerged, enabling the United Kingdom and British aircraft organisations to carve out a specialist niche market in the military world from 1919 to 1939. However, this approach did not favour the development of competitive commercial airline companies, which was also hampered by a lack of government funding in this area. The British Government's approach to funding aviation created a peculiar tension within many organisations, which led to interesting responses. Over the long run, the most successful firms were the ones that were able to achieve this balance and manage the tensions. Because of the technical dynamism of the industry, an inability to create new and better designs of aircraft usually led to company failure. This meant that in the early days of the industry, most of the firms formed were by the entrepreneurial pioneer aviators and builders of experimental aircraft. Understanding the nature and motivation of these founders can help understand the nature and motivation of entrepreneurs in high-tech industries.

In terms of the workforce structure, this largely changed in the interwar period. In each case, the top design engineers, factory directors and test pilots tended to dominate the firms and were held in high regard. There was also demand for a range of white- and blue-collar employees, arising from the needs of complex aviation production. This demand arose from research, design, planning, test piloting and other operations and the employment of structural engineers and aero dynamists, designers and draughtsmen, accountants, purchasing agents and a variety of other technicians and administrators. The white-collar staff made up a larger proportion than in other parts of the engineering industries. White-collar staff worked mostly in large design studios and laboratories or planning offices. Blue-collar workers typically comprised 60 to 80 per cent of the workforce. Many of these, however, were skilled, such as carpenters, metal fitters, welders, lathe operators, machinists and sheet metal workers. Skilled work teams were employed on the various parts of the planes before they were assembled.

In many of the smaller factories, engineers, technicians and tradesmen worked closely together, especially if working on prototypes. Originally, establishments tended to just be design offices attached to a small workshop, with the same being true of the engine makers, which had teams working on design, the building of prototypes and production. Skilled workers, therefore, made up a large proportion of the workforce. The emergence of the industry before the war and then the rapid growth of it meant that many foremen and managers were promoted from the ranks of technicians and tradesmen.

As time went on, shortages of wood encouraged some companies to experiment with metal construction and the resultant change towards the use of metal construction became one of the most important technical changes in the 1920s in aircraft manufacturing. The advantages of metal making use were to become very apparent. Metal structures could be checked scientifically for strength and consistency. It was better for large-scale production, and its life was longer. The introduction of metal construction meant the manufacture of stronger aircraft, which allowed for greater speeds to be achieved. The first all-metal, stressed-skin, civil monoplanes entered service in the United States in 1933 and within a few years, monoplanes had replaced biplanes. Production that made use of metals, however, was more expensive as special jigs and tools were necessary. The new construction methods also meant an increased commitment to testing and development, the acquisition of new machinery and modifications to manufacturing processes.

In addition to the use of metals, a range of other technological changes occurred. One of the most important was the development of considerably more powerful and lighter air-cooled radial engines. Other changes took place, such as the use of retractable landing gear and enclosed cockpits. Technical progress in other related areas took place and consisted of incremental innovations – e.g. the variable pitch propeller and slotted wing.

Bearing in mind these substantial technological changes in the industry, it is possible to use Sarasvathy's (2001) reasoning theory to see why the aviation industry was attractive to entrepreneurs. Aviation in its earliest days provided an opportunity for the entrepreneurs to develop their passions, rapidly develop their knowledge, experience and partnerships and proactively shape their careers. The demands of the early years of the aviation industry called for individuals with high levels of technical skill, which could be acquired in the fledgling aviation industry itself, or in related industries either through work experience, apprenticeships, or university or technical college engineering degrees/diplomas. The rapid growth of the industry, fueled by government investment in the development of military aircraft, provided significant career growth opportunities. Interesting and challenging work for individuals with technical capabilities were readily available and the tiny industry attracted dynamic and driven individuals (e.g. Glenn H. Curtiss, Glenn L. Martin, Frederick Handley Page, Thomas Sopwith, Robert Blackburn and A.V. Roe) who worked together in a tight and dynamic industry ecosystem. As the industry developed, these individuals rose quickly to become the heads of important companies.

Companies, to begin with, were able to manufacture aircraft mainly from wood with fairly modest organisations. This kept the start-up costs for new firms within reach for many of the pioneers, which encouraged them to start businesses in the industry. Nevil Shute Norway, later writing about this aspect of the industry, stated,

> In those days the capital requirement to start an aeroplane manufacturing company was not large, judged by modern standards. In 1930 the demand for small two- and three seater aeroplanes for personal and club flying was brisk all over the world. These units were small in value and since wooden construction was still the rule no very great numbers had to be produced to manufacture at a profit; the cost of jigs and tools for a given design was small by modern standards.
>
> (Shute, 1954, p. 104)

The post-World War I industry shakeout favoured those with an entrepreneurial flair. The industry led to two pathways – one to the production of specialised aviation products and the second to a diversification into other industries. Both pathways created significant creative opportunities for entrepreneurs, who generally shied away from bureaucratic management roles.

In many of the firms, the pioneering spirit and enthusiasm for aviation were strong enough amongst the managers such that they were to overcome the poor financial returns they received and retain an intense desire to remain independent. This meant that they focused on pursuing their personal vision, which made rationalisation of the industry into fewer, larger companies difficult to achieve. Some consolidation, however, eventually took place in the United Kingdom, the United States and France, encouraged by the increasing technological demands of the industry.

At the same time in the 1920s and 1930s, the governments of the United Kingdom, France and the United States all recognised a persistent threat of war and started to reinvest in the industry. To promote competition and production capacity, and to keep a reserve of design and production teams together, each government pursued the policy of spreading its military contracts over a number of firms. This meant that despite there only being a small market for aircraft at this time, there was an unusually large number of firms retained in each country that were manufacturing aircraft on a small scale. This helped to strengthen the ecosystem of the industry and allowed entrepreneurs to build capability by utilising their connections to move into areas where their passion lay; however, it also limited entrepreneurs starting up in the industry and also prevented the greatest economies of scale from being realised. The pressures to create economies of scale driven by technological change were, however, very real. Shute himself later commented on this change:

> As the swift years went by the capital required for profitable operation doubled and redoubled every year as aeroplanes grew larger and as wooden

> aircraft ceased to sell and had to be replaced by metal construction, but in 1930 forty thousand pounds seemed reasonable capital for such an enterprise.
> (Shute, 1954, p. 104)

What was true of the changes in the manufacture of airframes was also true in the other major part of the industry, the manufacture of aero engines. Unlike airframes, which were relatively easy to build, aero engines required more skilled labour and greater levels of capital. Airframe production was relatively easy to expand, as any firm used to working with wood could follow plans provided and begin production. The production of aero engines was more difficult. In France and Germany, the early establishment of the motorcar industry gave these two nations an early comparative advantage at developing aero engines, although the United Kingdom and the United States were both to develop advanced industries in this field later. The United Kingdom's early aero engine production problems were eventually solved by the motorcar companies Napier and Rolls-Royce, which first built French engines and then turned to developing their own. In the American case, in 1917, motorcar manufacturers produced the Liberty engine, an imported British design.

In the United Kingdom in the interwar period, engine manufacturing was largely undertaken by Napier, Rolls-Royce, Armstrong Siddeley and Bristol. Rolls-Royce was dependent on military sales, whereas Bristol was both a military and civil manufacturer of engines. Bristol tended to work on radial engines whereas Rolls-Royce built liquid-cooled designs. In the United States, the war left Curtiss Aeroplane and Motor and Wright Aeronautical as the principal manufacturers of aero engines. Until 1925, Wright was the only considerable producer of radial air-cooled engines, but in that year, Frederick B. Rentschler resigned as president of Wright Aeronautical in a disagreement over the company's policy towards manufacturing engines and organised a new company for the manufacture of radial air-cooled engines. This became the Pratt and Whitney Company, which specialised in the production of air-cooled radial engines. Similar specialisations took place in France, with the motorcar company Hispano-Suiza specialising in the production of liquid-cooled engines while the Gnome Company specialised in radials.

The industry in the United Kingdom, the United States and France in the 1920s and 1930s, therefore, was dominated by a number of relatively small- and medium-sized firms, each reliant on military sales. In the United Kingdom in 1934, for instance, the Air Ministry spent £6 million on military aircraft, exports (mostly military) earnt £1.5 million and home civil sales were £0.5 million. Even in the United States at this time, Army and Navy sales accounted for at least 50 per cent of sales between 1927 and 1933, rising to two-thirds in 1936. Export sales did rise, with American aircraft exports rising from under 10 per cent of production in the 1920s to over 40 per cent in the late 1930s (Edgerton, 1991, p. 18). Federal government expenditures for military aviation dropped from US$31 million in 1931 to US$13 million in 1933 but then rose to US$27 million by 1935 and US$70 million by 1939 (Rae, 1968, p. 110).

Rearmament from the mid-1930s onwards saw small-scale production begin to disappear, with production rising first in Germany and then the United Kingdom, France and, finally, the United States (see employment figures in Table 3.4; production figures in Table 3.5 and 3.8). This meant that the profitability of the aviation manufacturing companies also began to rise. Many of the companies made very little profits during the years of the 1920s and early 1930s and only did so from 1935 onwards (for an example of three British companies, see Table 3.9).

Table 3.8 Selected statistics on the aircraft production industry, 1924 to 1939

	United Kingdom			*United States*		
	Production no.	*Air Ministry*	*Military %*	*Production*	*Military*	*Military %*
1924	503	563	na	377	317	84
1925	na	448	na	789	447	57
1926	na	805	na	1,186	532	45
1927	na	392	na	1,995	621	31
1928	na	835	na	4,346	1,219	28
1929	na	615	na	6,193	677	11
1930	1,456	864	59	3,437	747	22
1931	na	na	na	2,800	812	29
1932	na	na	na	1,396	593	42
1933	633	na	na	1,324	466	35
1934	1,108	652	59	1,615	437	27
1935	1,807	893	49	1,710	459	27
1936	2,218	1,830	83	3,010	1,141	38
1937	2,827	2,218	79	3,773	949	25
1938	2,800	na	na	3,623	1,800	50
1939	7,940	na	na	5,856	2,195	38

*The number of aircraft ordered by the Air Ministry were higher than the number produced in 1924 because of the wartime stocks still not having been exhausted.

Source: Hayward (1989, p. 14), Edgerton (1991, p. 26), Modley and Cawley (1953), Truelle (1969)

Table 3.9 Profits of de Havilland, Handley Page and Hawker (Hawker Siddeley)

	de Havilland Aviation Co. Ltd.		*Hawker Aviation/ Hawker Siddeley*		*Handley Page Ltd.*	
	Net profits £	*Dividend paid on ordinary shares %*	*Net profits £*	*Dividend paid on ordinary shares %*	*Net profits £*	*Dividend paid on ordinary shares %*
1918	na	na	na	na	114,952	
1919	na	na	na	na	51,285	0
1920	na	na	na	na	−645,773	0
1921	2,387	na	na	na	−68,149	0
1922	4,467	10.0	na	na	3,767	0
1923		10.0	na	na	146,780	0

(*Continued*)

Table 3.9 (Continued)

	de Havilland Aviation Co. Ltd.		Hawker Aviation/ Hawker Siddeley		Handley Page Ltd.	
	Net profits £	Dividend paid on ordinary shares %	Net profits £	Dividend paid on ordinary shares %	Net profits £	Dividend paid on ordinary shares %
1924	8,531	10.0	na	na	−22,010	0
1925	14,306	10.0	na	na	38,982	0
1926	16,451	10.0	na	na	53,922	0
1927	18,350	10.0	na	na	46,638	0
1928	19,701	10.0	na	na	47,795	0
1929	48,710	10.0	na	na	40,508	0
1930	36,543	5.0	na	na	43,147	0
1931	27,729	5.0	na	na	3,594	0
1932	2,223	2.5	na	na	8,135	0
1933	63,439	7.5	na	na	20,441	0
1934	50,994	7.5	na	na	44,590	15
1935	77,367	7.5	195,801	na	77,893	15
1936	71,709	10.0	118,649	20.0	100,181	50
1937	122,184	12.5	362,443	27.5	210,126	20
1938	152,813	12.5	594,193	42.5	190,696	20
1939	195,217	12.5	757,767	42.5	219,479	20
1940	201,214	12.5	620,672	32.5	159,388	20
1941	102,714	10.0	536,122	32.5	148,383	25
1942	40,725	10.0	629,817	32.0	145,158	25
1943	52,937	10.0	529,444	32.5	145,176	25
1944	52,315	10.0	542,942	32.5	169,776	16.5
1945	49,215	10.0	245,511	32.5	169,100	10.0

Source: Fearon (1978). *The Scotsman*, 26 November 1936, p. 5, 16 November 1937, p. 4

Despite the slump in demand during the early 1930s, it was an important period in the aviation industry. In a brief two-year period in the early 1930s, streamlined, twin-engine, all-metal, low-wing monoplane bombers with retractable landing gears and enclosed cockpits were flown, tested and brought into operation in Germany, Russia, the United Kingdom, France and the United States. These were first used for commercial aircraft. In the United States, the employment of the bomber monoplane spread quickly. This bomber plane technology was dual use in that it could be used to develop commercial transport aircraft. But it was the Soviet I-16 that became the first operational, single-seat, low-wing monoplane fighter with retractable landing gears to fly in December 1933, three-and-a-half years after the Boeing Monomail Model 200 and more than six years after the first flight of the Lockheed Vega.

Besides military demand, the development of commercial aircraft was also an important part of the changes that occurred in the industry in the 1930s, even though it tended to be overshadowed by the use of aircraft for military purposes.

After a very unsteady start following the end of World War I, the commercial aircraft industry in the 1930s finally began to make progress, especially in the United States.

3.5 Commercial flying

Before World War I, there had been no regular air services. This changed in the 1920s, but early air transport companies faced a number of problems, the most serious being that none of them had the time or money to develop aircraft specifically designed for commercial work. Other difficulties existed, including the lack of ground organisations and suitable aerodromes, adverse weather conditions and a general lack of public interest in travelling by air.

As commercial markets slowly developed in the 1920s new technologies and designs were gradually pioneered, generally after they had been used in the production of military aircraft. To the development of the commercial industry, the ten-year period from 1925 to 1935 was particularly important, as a series of advances in aeronautical engineering and aerodynamics occurred that was to revolutionise commercial aviation. These changes meant that it became possible to design and build passenger transport aircraft whose operating costs and payload were such that they made possible services that were unsubsidised. These changes occurred first in the United States, where competition between airlines helped to create a demand for higher performance aircraft with low operating costs. This, in turn, helped to hasten the development of monoplane airliners. The search for a truly economic airliner to feed the growing demand for air travel was a dynamic stimulus during the 1920s and 1930s.

The geography of the United States, with huge continental distances between population centres, was an ideal context for air travel. The United States Government also encouraged the growth of civil aviation through mail contracts after 1925. Local and state governments also provided assistance by investing in new and enlarged airports.

Moreover, competition between several carriers put a premium on sound and efficient equipment and formed a sufficiently broad base to support a number of producers. Once established, the level of demand provided major manufacturers with returns which justified further investment in large-scale, modern production facilities and a commitment to research and development.

To see the advantage that the American industry had, Table 3.10 provides data on the number of passenger kilometres travelled in a range of countries in 1929 and 1938. As can be seen from the figures, the American market was far larger than that of the European one. What can also be seen is that the German industry was the largest in Europe, even in the 1920s when it was banned by the Versailles Treaty from operating an air force. Early in the 1920s, the air transport companies in France were heavily government subsidised and could, consequently, have the advantage of being able to offer cheaper fares. This came mainly at the expense of the British companies. Finally, in 1924, the British Government organised the creation of a government-backed civil aviation monopoly, Imperial Airways, which absorbed the existing air transport operations. This new entity was subsidised, but to a lesser degree than its German and French counterparts. In 1927–28, for instance, the British Government's expenditure on the industry was US$230,000 compared to Germanys US$1.2 million and France's US$4.5 million (Dobson, 1991, p. 43). Despite the scale of the French subsidies, the French industry gradually fell behind that of both Germany and the United States (see Table 3.10). In the German case, the various German airlines were merged in 1926 to form Deutsche Lufthansa, which by 1939 was Europe's largest airline.

Despite this growth of the European commercial aviation industry, military demand for aircraft was still far in excess of commercial demand. Advances in American technology meant that in the 1930s European airlines began purchasing American commercial aircraft, such as the Dutch airline KLM's purchase and use of DC3s. By this stage, European manufacturers were still relatively well advanced in the construction of military aircraft but had lagged behind that of the American producers in the production of commercial aircraft.

The outbreak of World War II helped to create an additional boost in demand that greatly increased the scale of the industry. By the end of this war, the industry was once again entering a new phase of technological innovation with the development of jet aircraft, which, along with the development of commercial air travel and Cold War demand for military aircraft, made the transition to peacetime production a relatively easy one compared to the post-World War I industry. The industry by this time, however, had become a very different one to the one that had existed in the early 1920s. Economies of scale, high capital intensity and

Table 3.10 Passenger travel, million passenger kilometres

	1929	*1938*
Belgium	1.6	18.0
Czechoslovakia	0	22.0
Denmark	0.5	5.6
Finland	0.8	3.4
France	12.0	73.0
Germany	24.0	128
Ireland	0	2.0
Italy	40.0	209
Netherlands	7.1	60.0
Norway	0	2.4
Poland	3.8	9.4
Spain	2.1	11.0
Sweden	1.2	14.0
Switzerland	2.4	11.0
United Kingdom	11.0	86.0
Yugoslavia	0	2.0
Total Europe	106.5	656.8
Canada	9.8	35.0
United States		1,215.0

Source: Mitchell (2007, 2008)

highly educated research teams dominated the profile of the leading companies. In a relatively short period of time, the industry had made a considerable transition from a small-scale cottage industry into an industrial leviathan.

3.6 Conclusion

The aviation industry began in the 1900s as one based on handicraft skills, with firms building a small number of planes by enthusiastic pioneer aviators for wealthy sportsmen. Most early participants had engineering backgrounds and skills, and there was a close association between the aviation, motorcycle and car industries, especially in terms of the development of aero engines, but also in the skill training of many of the early aviators. Aero clubs and exhibitions all helped to fund and spread the new technology, with the military of each country beginning to take an interest in the development of the industry pre-World War I.

World War I quickly changed the industry into one of mass, industrial-scale production, with technological changes being introduced at a rapid rate. These technological changes continued after the war and by the 1930s had enabled a commercial passenger and freight industry to develop. The period up until the end of World War II was one of constant technological improvement, and although the late 1920s and early 1930s were years of particular importance, even in other years, technological improvement continued.

In the interwar period before 1935, military demand tended to be limited, but the governments of France, the United Kingdom and the United States spread contracts around manufacturers in order to keep design teams and production capacity intact. The production of aero engines, was, however, more concentrated and limited to a small number of companies in each country, mainly because of the more complex nature of aero engine production. In each country, motorcar companies were important in the development of aero engines.

By the late 1920s, the industry had evolved into a far more complex one than it had been in the pre-World War I period. Despite this production, units remained relatively small, reflecting the considerable level and diversity of skill required for effective design teams and production workforces.

On the demand side, there was concentration as well, which came first mainly from the military and then from a relatively small number of commercial transport companies. Even with greatly reduced budgets of the interwar years, military purchases tended to dominate the industry, and only in the United States was this contested by the development of commercial passenger aviation. Private flying still occurred over the years but was harshly affected by the Depression in the early 1930s. From 1935 onwards rearmament dominated the industry, and production in all countries grew considerably in size.

Through all of these changes, a few key pioneer aviators, innovators and entrepreneurs survived the dramatic changes to the industry, designing and flying new models of aircraft attracted to the continued excitement of this dynamic industry. This was true in the United States, the United Kingdom and France, and included many of the great names of early aviation. It's to these people that we now turn more specifically.

Note

1 The Wright brothers are generally given credit for being the first to fly powered, heavier-than-air planes. Even so, a number of claimants exist in a range of countries; for instance, it was asserted that Richard Pierce flew an aircraft in New Zealand in early 1903.

References

Angelucci, E. (2001). *Illustrated Encyclopaedia of Military Aircraft, 1914 to the Present*. Edison, NJ: Chartwell Books.

Bilstein, R. E. (2001). *The Enterprise of Flight: The American Aviation and Aerospace Industry*. Washington, DC: Smithsonian Institute.

Bishop, C. (2002). *The Encyclopaedia of Weapons of World War II*. London: Sterling Publishing.

Chadeau, E. (1986). *De Blériot a Dassault: Historie De L'Industrie Aeronautique En Framce 1900–1950*. Paris: Fayard.

Dobson, A. P. (1991). *Peaceful Air Warfare: The United States, Britain and the Politics of International Aviation*. Oxford, UK: Clarendon Press.

Edgerton, D. (1991). *England and the Aeroplane: An Essay on a Militant and Technological Nation Centre for the History of Science, Technology and Medicine*. Basingstoke, UK: Macmillan Publishing Limited.

Fearon, P. (1969). The formative years of the British aircraft industry, 1913–1924. *Business History Review*, 43, pp. 476–495.

Fearon, P. (1974). The British airframe industry and the State, 1918–1935. *Economic History Review*, 27(2), pp. 236–251.

Page, F. H. (1944). The future of the skyways: A British view. *Foreign Affairs,* 22(3), pp. 404–412.

Hayward, K. (1989). The British aircraft industry. Manchester, UK: Manchester University Press.

Higham, R. (2012). *Two Roads to War: The French and British air Arms from Versailles to Dunkirk*. Annapolis, MD: Naval Institute Press.

Mitchell, B. R. (2007). *International Historical Statistics: The Americas 1750–2005*. Basingstoke, UK: Palgrave Macmillan.

Mitchell, B. R. (2008). *International Historical Statistics: Europe 1750–2005*. Basingstoke, UK: Palgrave Macmillan.

Modley, R. and Cawley, T. J. (1953). *Aviation Facts and Figures 1953*. Washington, DC: Lincoln Press.

Owen, G. (1999). *The Decline and Revival of British Industry Since the Second World War*. London: HarperCollins.

Rae, J. B. (1968). *Climb to Greatness: The American Aircraft Industry 1920–1960*. Cambridge, MA: MIT Press.

Roe, A. V. (1939). *The World of Wings and Things*. London: Hurst & Blackett.

Sarasvathy, S. D. (2001). *What Makes Entrepreneurs Entrepreneurial?* Charlottesville, VA: Darden Business Publishing.

Shute, N. (1954). *Slide Rule: The Autobiography of an Engineer*. London: William Heinemann Ltd.

Truelle, J. (1969). La production aeronautique militarie Francaise Jusqu'en Juin 1940. *RHGGM*, 73, pp. 89–110.

United States, Bureau of Census. (1927). *Census*. Washington, DC.

Van der Meulen, J. A. (1991). *The Politics of Aircraft: Building an American Military Industry*, Lawrence, KS: University Press of Kansas.

Part II

Entrepreneurial aviators

4 The earliest days in the United States

The Wright brothers, Glenn H. Curtiss, Allan Lockheed and Glenn L. Martin

4.1 Introduction

Although the Wright brothers are best known as the originators of powered, heavier-than-air flight, they were also the founders of a company that manufactured aircraft. Other pioneers of the pre-World War I era, such as Allan Lockheed, Glenn L. Martin and Glenn H. Curtiss, also established companies to manufacture aircraft at this time. The Wrights' company eventually merged with Curtiss' and the Curtiss-Wright Company is still a part of the aerospace industry today. The creation of these various companies by the pioneers of flight is a solid indication that these pioneers of flight in America were not just interested in the immediate sensation of flight, but they were also interested in realising a longer-term vision of aviation that incorporated the use of aircraft in commercial activities.

In the period between the Wrights' first flight at Kitty Hawk in 1903 (see Figure 4.1) and the outbreak of World War I, the market for aircraft in the United States was very limited. Despite this difficulty, the Wright brothers and others put a great deal of effort into creating various business enterprises, the purpose of which was to manufacture and to sell aircraft to the American public.

The industry itself, besides the commercial attractions, had a powerful non-economic attraction for its participants and youthful spirits, who had an eye for the extraordinary. In the early days, the industry was a knowledge- and craft-based activity, and aircraft manufacture was not regulated or structured. This meant substantial start-up capital was not required, offering an open space for new entrants to design and experiment. The potential for the future was recognised by a few entrepreneurs who pursued its realisation.

In this period the American industry consisted of a series of small workshops scattered across the nation, operated by pioneer aviators who experimented and constructed a handful of aircraft each year for wealthy sportsmen. Design risks and the loss of life were high and revenue streams were not established, which dissuaded non-entrepreneurs and established businesses from entering the market.

The military at the same time showed a lack of interest, and it was only under pressure from President Theodore Roosevelt that an apathetic War Department made its first call for bids on aircraft, resulting in a US$30,000 contract being signed with the Wrights in February 1908. Beginning in 1909, small but regular

Figure 4.1 Seconds into the first flight at Kitty Hawk, North Carolina, 17 December 1903

appropriations were made for Army and Navy aviation. These grew more substantial after the outbreak of World War I in 1914, and by the end of 1916, there were at least 14 firms employing 10,000 workers competing for US and foreign contracts. Two firms – the Curtiss Aeroplane and Motor Company and Wright-Martin Corporation – were capitalised at US$10 million each (Vander Meulen, 1991, p. 12).

Despite the very small size of these early endeavours, a number of the United States' greatest aerospace companies, such as Curtiss-Wright and Lockheed Martin, trace their origins back to these tiny concerns. The purpose of this chapter, therefore, is to look at a small number of the most important of these earlier aviation pioneers and business entrepreneurs and their efforts to create an industry virtually from scratch.

4.2 The Wright brothers

The Wright brothers (Orville Wright, 1871–1948 and Wilbur Wright, 1867–1912) were two of America's earliest aviation pioneers, but they were also engineers and inventors. They are generally credited with having invented, built and flown the world's first successful aircraft. The brothers made their first controlled and sustained flight of a powered, heavier-than-air aircraft on 17 December 1903 at a location four miles south of Kitty Hawk in North Carolina. Subsequently during 1904 and 1905, the two brothers worked on their flying machines, developing them into the first practical fixed-wing aircraft (Anderson, 2004; Russell, 1974; Crouch, 2003; Howard, 1988; McPherson and Gardner, 2004; Tobin, 2004).

Wilbur was the eldest of the two and was born near Millville, Indiana, in 1867. Orville was born later at Dayton, Ohio, in 1871. The brothers never married and were part of a family of seven children (two of whom died in infancy). Their father was a bishop in the Church of the United Brethren in Christ and both brothers attended high school but did not complete their schooling or receive diplomas.

After their mother died in 1889, the household of which they were a part was run by their older sister, Katharine, a school teacher and proactive advocate of the right of women to vote. In the same year that their mother died, Orville dropped out of high school and began to run a printing business, having designed and built his own printing press with Wilbur's assistance. Wilbur also left school later without completing it and joined Wilbur in the printing business. In December 1892, the two brothers opened a bicycle repair and sales shop (the Wright Cycle Exchange, later the Wright Cycle Company), having been attracted to the national bicycle craze which was occurring at the time, which in turn was spurred on by the invention of the safety bicycle and its considerable advantages over the penny-farthing. This led in 1896 to the Wright brothers beginning to manufacture their own bicycles, and they later used this business endeavour to fund their growing interest in flight. Experience in the construction of bicycles also helped them in the preliminary construction of aircraft. This construction work was also stimulated by their reading about the efforts of the German glider expert, Otto Lilienthal.

The Wrights gained the mechanical skills that were essential to their success in building the early aircraft by working for years in their own bike shop, where as well as working on bicycles they also repaired a range of other mechanical devices such as printing presses, motors and other machinery. Their work with bicycles, in particular, influenced the Wrights' view that an unstable vehicle like a flying machine could be balanced and controlled with practice. In undertaking their experimental work from 1900 until their first powered flights in late 1903, they conducted extensive tests with gliders, which helped to develop their skills as pilots. In doing so they came to regard control of the aircraft in flight as being the crucial unsolved third part of the problems associated with beginning flight. They believed that knowledge of the other two issues (engines and wings) was adequate for achieving flight (Crouch, 2003, p. 166). The brothers in 1900 first travelled to Kitty Hawk to begin their manned gliding experiments. This place was chosen for these experiments as it had regular mid-Atlantic coastal breezes, a soft, sandy landing surface, and was secluded.

The brothers in 1903 built the powered Wright Flyer I, which they were to use in their first flights on 17 December 1903. These early flights were followed in 1904 with flights using the newly constructed and improved Flyer II at Huffman Prairie, northeast of Dayton. In 1905 they built and flew a third new aircraft, the Flyer III. Over this period the brothers were able to achieve their goal of creating a practical flying machine, and after improvements had been made, they began to attempt to discover a means to commercialise their discoveries.

At first, news of the Wrights brothers' early flights were not widely known outside of the state of Ohio and were often met with some scepticism (Gollin, 1984, pp. 66–68). The brothers themselves were partially responsible for the lack of attention they received, as they had a tendency to attempt to guard their secrets. This was due to a fear that competing inventors might try to steal their ideas. In order to make some sales without attracting too much attention, they approached the military, first writing to the United States Government and then after getting no response, approaching the governments of France, the United Kingdom and Germany. In responding to the Wright brothers' letters, the United States military expressed no interest in their claims, and it was then that the brothers turned their attention to Europe. In each case, they had no success, a result that was largely a consequence of them insisting on a signed contract without undertaking demonstration flights. Their insistence on this point was a result of their desire for secrecy. This secrecy meant that through most of the 1900s, the European newspapers, particularly those in France, were openly dismissive of their claims.

The Wright brothers attempted no flights at all in the years 1906 and 1907 but instead spent their time trying to persuade the aforementioned governments that they had invented a successful flying machine. The country that especially attracted them was France, because of the high levels of general enthusiasm for aviation. Finally, they travelled to France in 1907 for talks with government officials and businessmen. In addition, the Wrights also met with aviation representatives in the United Kingdom and Germany. The Wrights, therefore, were doing business in Europe with interested governments for a year before the United States became interested.

Sceptical as they were, the French insisted on demonstrations, and Wilbur began their first official public demonstrations on 8 August 1908 at the Hunaudières horse racing track near the town of Le Mans in France. In the days that followed, Wilbur made a series of technically challenging flights, demonstrating to the French crowds his skills as a pilot and the capability of the Wrights' flying machine. In doing so he far surpassed those of all other pioneering aircraft and pilots of the day. The French public was enthralled by Wilbur's feats and crowds flocked to see him by the thousands, which made the Wright brothers instantly famous. Orville followed his brother's success by demonstrating another nearly identical Flyer aircraft to the United States Army at Fort Myer, Virginia, starting on 3 September 1908. On 17 September, however, Orville was seriously injured in a flying accident.

The success of these demonstrations meant that orders for planes began to come in. The Wright brothers were well placed in Europe to produce their aircraft, having built a factory in Germany (Flugmachine Wright Gesellshaft) in May 1909. This was around the time they also licenced the Short Brothers in England to build a dozen Wright Flyers. During their European trip in August 1909, they toured the Short Brothers Factory with Charles Rolls. A contribution to the success of the European trip was the addition of the brother's sister Katharine, who charmed the social scene, in contrast to her shyer brothers. Upon their return to the United States,

Katharine took on some of the business responsibilities of the Wrights' enterprise, becoming an officer of the Wright Company in 1912 and remaining there until Wilbur's death.

The Wright Company was incorporated in America on 22 November 1909 and the Wright brothers sold their patents to the American company for US$100,000. In doing so, they received one-third of the shares in a million dollar stock issue and a 10 per cent royalty on every aircraft sold (Crouch, 2003, p. 41). Wilbur became president of the company and Orville the vice president. They set up a factory in Dayton and a flying school/test flight field at Huffman Prairie. They did not, however, locate the headquarters of the company in Ohio but instead established in New York City, where many of the initial financiers of the enterprise were located. The company initially concentrated on building small numbers of aircraft, developing new models and running a training school for military and civilian flyers.

Unlike many of the other pioneer aviators, the sensational nature of the exhibition flights meant that the Wrights attracted the attention of a number of important financiers and therefore the capital raised for this venture was sufficient to put the new company on sound footing. In terms of financing, the Wright brothers were therefore to some extent an exception to the early experiences of most of the pioneers who tried to establish aircraft manufacturing companies. Their research and development was carried out with the earnings of their bicycle shop, but once they attracted public attention, they found interest from a number of financiers. Godfrey and Samuel Cabot of Boston offered to assist them as early as 1906 and the syndicate that formed, the Wright Company (in 1909), included a number of leading financiers of the time including Cornelius Vanderbilt, August Belmont and Russell and Frederick Alger (Rae, 1968, p. 100).

The capabilities of the Wrights, however, were not able to match this support. In writing about their efforts, Roach (2014, p. 2) argued that they were excellent self-taught engineers who could run a small company like the bicycle shop, but they did not have the business skills or temperament to run a substantial venture in an industry like aviation.

Despite these managerial deficiencies, during 1910 the Wright Company hired a workforce and moved into the new factory in Dayton. The main problem they encountered, however, was that despite their making successful demonstrations of flight, there were simply not many customers for aircraft. In the spring of 1910, therefore, the Wright brothers hired and trained a team of salaried exhibition pilots, who then showed off their machines and won prize money for the company. This was one of the few steady sources of income in the early days of flight, one that was turned to by others such as Curtiss, who also came to rely on aerial exhibitions. In addition, to keep the company going, the flying school trained 115 pilots who were instructed by Orville and his assistants. This was not a novel approach as many early manufacturers of aircraft ran flying schools on the side to raise money. In terms of manufacturing aircraft, as late as 1915, the company had only sold 15 aircraft to the United States Army, and civil sales limped along. The Aeronautical Division of the United States Army Signal Corp received its first military aircraft, a Wright Model B, in 1909; however, it remained the Army's sole aircraft for two years.

Despite all the publicity received after 1908, the two brothers didn't end up running the company for very long. Wilbur, the more capable administrator of the two, gradually became occupied with dealing with different lawsuits to maintain the patents the company had on aircraft. He became ill on a business trip to Boston in April 1912 and subsequently died from typhoid fever. Upon Wilbur's death, Katharine stepped down from her involvement in the business, and Orville subsequently estranged himself from Katharine when she became involved with and married a Kansas City newspaper owner and editor. This rift existed until her death in 1929.

Orville succeeded to the presidency of the Wright Company upon Wilbur's death, but he did not possess his brother's or sister's executive skills. Eventually, Orville sold the company in 1915 after buying up the company's shares in the previous year and offering the entire company for sale. In the autumn of 1915, a syndicate of New York financiers paid an estimated US$1.5 million for the company plus a retainer of US$25,000 to claim Orville's services for a year. After Orville left the company it became less focused on the manufacture of aircraft and by the 1920s it mainly produced aero engines for other aircraft manufacturers. In 1929, the Wright Company merged with the Curtiss Company.

Orville subsequently moved into other fields, making his last flight as a pilot in 1918. He retired from business and became an elder statesman of aviation, serving on various official boards and committees, including the National Advisory Committee for Aeronautics, the predecessor agency to the National Aeronautics and Space Administration (NASA), and the Aeronautical Chamber of Commerce, predecessor to the Aerospace Industries Association. He died on 30 January 1948, over 35 years after his older brother.

In their years together as businessmen, the two brothers tended, after their early entrepreneurial endeavours, to concentrate on the protection of their intellectual property as much as anything. They made their first United States patent in 1906, which did not claim invention of a flying machine, but instead the development of a system of aerodynamic control that manipulated a flying machine's surfaces (a wing-warping technique that gave an aircraft lateral stability). It was this innovation that more than anything made it possible for the brothers to take to the air. The Wright's main legal struggle was with Glenn H. Curtiss, who with Alexander Graham Bell had developed and patented the aileron (the trailing edge of a plane wing) in 1911.

After Curtiss made an important flight of one kilometre on 4 July 1908in his Aerial Experiment Association (AEA) June Bug plane, the Wrights warned him not to infringe their patent by profiting from flying or selling aircraft that used ailerons. Curtiss refused to pay the Wrights licence fees and in 1909 sold an aircraft equipped with ailerons to the Aeronautic Society of New York. The Wright brothers responded by filing a lawsuit, which began a long legal conflict. In addition, they also sued overseas aviators who flew at United States exhibitions. The Wright brothers won their initial case against Curtiss in

February 1913, when a judge ruled that ailerons were covered under the patent; however, the Curtiss Company appealed the decision, and in January 1914, a United States Circuit Court of Appeals upheld the verdict against the Curtiss Company, which continued to avoid penalties through legal tactics. For his part Curtiss was not reluctant to attempt to legally protect his innovations, defending claims to some 30 patents on seaplanes in the courts. The patent issue was not finally dealt with until after the entry of America into World War I and after the Wrights themselves had left the industry. During the war, pressure by the United States Government led to the creation of the Aircraft Manufacturers' Association and a cross-licencing agreement, whereby the member companies paid a blanket fee for the use of aviation patents, including the original and subsequent Wright patents. The Wright-Martin Company (successor to the Wright Company) and the Curtiss Company (which held a number of its own patents) each received a US$2 million payment from this agreement. The 'patent war', therefore, came to an end, although side issues lingered in the American courts until the 1920s.

After Orville sold the company, efforts were sought by the new owners to merge the company with others. The company at this time concentrated in engine manufactures, with the investors buying an interest in the Simplex Automobile Company, and built Hispano-Suiza engines in the United States under licence from the French Government. The company produced around 5,000 engines during the war.

The Wright Company initially merged with the Glenn L. Martin Company of Los Angeles, but Martin left the merged firm to revive the Glenn L. Martin Company. Wright-Martin, in turn, was dissolved in 1919 and reorganised as Wright Aeronautical. A new group of investors joined the company, and Frederick Rentschler, an engineer who had been involved in producing the Hispano-Suiza engine, became vice president and general manager, and in 1921, he became president. In a twist of irony, the Wright Aeronautical Corporation (another successor) and the Curtiss Aeroplane Company merged in 1929 to form the Curtiss-Wright Corporation, which remains in business today producing high-tech components for the aerospace industry. By this time, the various aviation pioneers had left the companies that they had originated, allowing for the merger to occur without opposition from them.

The Wrights had a rather checkered career as aviation manufacturers. Neither had the necessary skill or experience to manage a substantial aviation manufacturing concern, and even in the case of their technical abilities seem to have been more focused on legally protecting their own early inventions rather than developing new ones. By 1911, Wright aircraft were considered inferior to those of European makers, and indeed, aviation development in the United States was so slow at this time that when the country needed planes, no acceptable American-designed aircraft were available, and instead, American forces were compelled to use French machines. Even in the United States, the Wrights lagged behind others, and when the American military began buying additional aircraft after the outbreak of World War I, it tended to prefer Curtiss aircraft to Wright ones. This tendency explains the Wright Company's growing reliance on the sale of aero engines at this time.

In the Wright businesses, Orville was the starter, and Wilbur and Katharine were the runners, with the former doing the innovation, while the latter two looked after the administrative side. Their approach was to develop and then establish dominance over the fermentation stage of the industry by initially not demonstrating sustainable flight and then controlling through patents their system of aerodynamic control, a key aspect of sustained flight. The problem for the brothers was that the market was still emerging and their initial reluctance to demonstrate their technology did nothing to build credibility and trust in the new technology. A lack of orders must have emphasised to the Wright brothers that without the demonstration of sustained flight they would get no traction in the market – hence the 1908 European flight that kick-started the industry. They then sought to control the American and European markets through their patents – successfully in America, less successfully in Europe. The European flights attracted financing from backers who were early adopters, but Wilbur's death and Katharine's simultaneous departure from the company meant Orville was unable to deliver on the administrative requirements of a growing organisation.

The introduction of major shareholders into the company needed Wilbur and Katharine's administrative abilities as the family moved away from Orville's innovation to selling planes and protecting their patents. Unsurprisingly, Orville quickly sold the business after Wilbur's death to others with access to the administrative capabilities needed to develop the resources for running a growing manufacturing organisation. It appears that the Wrights were unable or unmotivated to keep up with the fast changes in the emerging market after Wilbur's death, preferring instead to harvest the patent royalties.

By examining the Wrights' collective career, we can see how the brothers shaped their aviation interests, first through experimentation and then by establishing and controlling key patents necessary for the development of aviation at the time. Their careers incorporated several companies and jobs, but all these lay within their controlling interests. Barriers to their progress were overcome in a pragmatic way – strategic demonstrations in Europe and the United States to key potential investors like the military and governments to overcome poor early sales. It could also be argued that as their influence developed, their relationships with others became strategic and focused on facilitating the realisation of their interests, demonstrating causation logic. It could be argued that Wilbur's death and Katharine's estrangement were dramatic changes, which Orville did not fully overcome, leading ultimately to the Wright Company moving out of family control.

4.3 Glenn Hammond Curtiss

The company that the Wrights' company was to eventually merge with was that of their great rival Glenn H. Curtiss (1878–1930) (see Figure 4.2). Curtiss was born in 1878 in Hammondsport, New York, and although his formal education

Figure 4.2 Glenn H. Curtiss

only went as far as the eighth grade, he gained considerable experience working in mechanical and engineering roles as a youngster. His early interest in mechanics and inventions was evident at his first employment which was with the Eastman Dry Plate and Film Company (later Eastman Kodak Company) in Rochester, New York. Curtiss was a compulsive inventor of new products and at an early age invented a stencil machine, which was adopted by his employer, and later built a rudimentary camera which he used to study photography. Unlike the Wright bachelors, he married in 1898 and had two children.

Curtiss subsequently worked in a variety of jobs including as a Western Union bicycle messenger, a bicycle racer and bicycle shop owner, and in 1901, he developed an interest in motorcycles when internal-combustion engines became more common. In 1902, Curtiss began building motorcycles with his own single-cylinder engines that he designed, and in 1903, he set a motorcycle land speed record. Still later on 21 January 1906, he set an unofficial world record of 136.36 miles per hour, riding a motorcycle he designed and built himself, and it was his success at racing that built him a reputation as a leading maker of high-performance motorcycles and engines. It was from this beginning that he turned to aviation and as early as 1904, he began to build airship engines (Casey, 1981; Charles and House, 2001; Roseberry, 1972; Shulman, 2002; Hatch, 2007).

Because of his reputation as an engine builder, in 1907 Alexander Graham Bell invited Curtiss to design and build an engine for experiments in heavier-than-air flight. At this stage, Bell regarded Curtiss as 'the greatest motor expert in the country' and invited him to become a member of his AEA. This association was a pioneering research group and between 1908 and 1910, it was responsible for building four aircraft. Curtiss primarily designed the AEA's third aircraft, Aerodrome #3, and afterwards was its test pilot. These flights were regarded as being the first pre-announced public flights of a heavier-than-air flying machine in the United States, and Curtiss afterwards went to Europe and won a race at the world's first international air meet in France. This meet took place in August 1909 at the previously mentioned Grande Semaine d'Aviation aviation meeting at Reims, France, organised by the Aéro-Club de France. The Wrights didn't compete in this event, as they were selling aircraft in Germany at the time, although two Wright aircraft were flown at the meet, but they did not win any events. Curtiss flew his own No. 2 biplane and won the overall speed event, beating Louis Blériot, who came in second. Back in the United States, Curtiss, on 29 May 1910, flew from Albany to New York City, making the first long-distance flight between two major cities in that country. He completed this 137-mile flight in four hours and in doing so won a US$10,000 prize, which had been offered by publisher Joseph Pulitzer. He was also was awarded permanent possession of the *Scientific American* trophy. Effectively, therefore, Curtiss began doing what was common amongst the early aviation pioneers. He had a love for speed in the form of bicycle and motorcycle racing. He also used the publicity of record-breaking flights and the winning of prize money to help start his business ventures.

To this end, Curtiss formed the Herring-Curtiss Company with Augustus Moore Herring on 20 March 1909. This company was subsequently renamed the Curtiss Aeroplane Company in December 1910, as his partner by then had left the company. The Herring-Curtiss Manufacturing Company was capitalised at US$160,000, and it was intended to build aircraft as well as other motor driven vehicles and engines at Curtiss' Hammondsport factory. The stock was divided into 1,800 shares of preferred stock and 1,800 of common stock, each valued at US$100. The principal stockholders of Curtiss' company were long associates, along with other (mainly New York–based) businessmen, with Curtiss receiving 160 shares of preferred and 400 shares of common stock, although Herring was the majority stockholder with 2,000 shares of preferred and common stock. Curtiss assigned all the assets of the old company, with his patents and real estate holdings and Herring contributed patents, some shop equipment and US$650. Curtiss and Herring each received US$5,000 a year in salaries, Monroe Wheeler was elected chair of the board of directors, Curtiss vice president and manager, and Herring second vice president (Trimble, 2013, p. 54; *New York Times* 21 March 1909). The company, however, was declared bankrupt on 1 April 1910, and Curtiss bought back his Hammondsport factory from the trustees handling the bankruptcy (Trimble, 2013, p. 81; Roseberry, 1972, pp. 253–254).

A new company was formed on 19 December 1911 as the Curtiss Motor Co., with the Curtiss Aeroplane Co. as a subsidiary. Curtiss was president and general manager, and held half of the US$600,000 in capital. The motor company stayed in Hammondsport while the aeroplane company shifted to Buffalo in 1915. The Curtiss Exhibition Company, which demonstrated Curtiss planes at events, was also founded at this time, and a subsidiary, Curtiss Aeroplane and Motors Ltd., was later organised in Canada in February 1915 to supply planes and plants for the British from a factory in Toronto.

The real support for Curtiss and his company once it was established was from his own business interests in the manufacture of motorcycle engines, along with the prize money that he had won. Later, the company was reformed on a larger scale when in early in 1916, the Curtiss Aeroplane and Motor Corporation was created by combining the Curtiss Company with the Burgess Aeroplane Company.

Curtiss' company concentrated on building aircraft for the United States Army and Navy, and during the immediate pre-World War I years, his experiments with seaplanes led to a number of advancements in naval aviation technology. One of these took place on 14 November 1910 when the Curtiss demonstration pilot, Eugene Ely, flew off from a temporary platform that had been mounted on the forward deck of the cruiser USS *Birmingham*. This successful takeoff and the subsequent flight to shore began the long and enduring relationship between the Navy and Curtiss. This was further enhanced at the end of 1910 when Curtiss set up a winter encampment at San Diego where both Navy and Army personnel were taught to fly. A few years later, B. Douglas Thomas, formerly of Sopwith Aviation in England, was recruited by Curtiss in the spring of 1914 to help with the development of new Army trainers.

Like the Wright brothers in 1909 and 1910, Curtiss employed a number of demonstration pilots, which were used to promote this aircraft. At this time, aerial competitions and demonstration flights helped to promote aviation by introducing to aviation to a fascinated public. Curtiss used these occasions as best he could to promote the sale of his aircraft.

The late American entry into World War I, however, meant that the industry in that country had a much shorter period of time in which to respond to the high procurement demands of the war. Even before the war, the American industry had lost its technical leadership, with this loss in part being due to the contentious litigation battle over patent rights that raged between the Wright brothers and Curtiss, as well as to a range of other factors. Despite the late entry into the war, Curtiss was able to take advantage of the growth of European demand after the start of World War I, with the Royal Navy purchasing several models of seaplanes from him. The Curtiss factory manufactured and sold to the British 68 'Large Americas', which later evolved into the H-12, the only American-designed and American-built aircraft to see combat in World War I.

In 1916, as America's entry into the war approached, the United States Army's Aviation Section, the United States Signal Corps, became interested in financing the development of a simple, easy-to-fly-and-maintain two-seat trainer. In response, Curtiss developed the JN-4 Jenny for the Army and the N-9 seaplane version for the Navy. These two planes were amongst the most important ever developed by Curtiss and thousands were built and sold to the militaries of the United Kingdom, Canada and the United States. During the war civilian and military aircraft demand boomed, and the Curtiss Company ended up employing 18,000 workers in Buffalo and 3,000 workers in Hammondsport. The Jenny was the only indigenous design manufactured during the war by the American aviation industry in large numbers, with most manufacturers choosing to copy English and French types. In addition, in 1917, the United States Navy commissioned Curtiss to design a long-range, four-engine flying boat which would be large enough to hold a crew of five. This he did and it was known as the Curtiss NC.

Curtiss' expansion into wartime production took place at a difficult time for the infant American aviation industry. The industry expanded output in response to the lead up to war, but remained modest in size. In the early years of the war, the European lead widened, with the European powers experimenting with a wide variety of airframe and engine configurations, and with large scales of production and turnover. When the United States finally entered the war in April 1917, the country had developed no mass-produced, modern fighter or bomber aircraft. Production expansion was, therefore, very ad hoc and made up largely by motorcar companies. The two pre-war aviation companies that were important at this time were Wright-Martin and the Curtiss Company.

The increase in demand for Curtiss aircraft in the run-up to the United States entry into the war put pressure on Glenn H. Curtiss' management of his company. On 13 January 1916, the acquisition of the Curtiss companies by a Wall Street combine occurred. This was headed by a New York investment firm, which folded Curtiss Aeroplane, Curtiss Motor, Canadian Curtiss Aeroplanes and Motors and Curtiss Exhibition into a new Curtiss Aeroplane and Motor Co. capitalised at US$9 million. Curtiss became president and received US$5 million in cash and US$3.5 million in preferred stock (Trimble, 2013, p. 182; Roseberry, 1972, pp. 402, 403, 406; *New York Times* 29 June, 14 July 1917). On 14 July 1917, John N. Willys succeeded Curtiss as president, and he sought to apply automotive mass-production expertise and efficiency. Curtiss became chairman of the board of directors, and he concentrated on experimentation and research as head of a subsidiary, the Curtiss Engineering Corporation, created a month later. This arrangement meant that he effectively left the management of the mass production of aircraft to others more suited to it, while he concentrated on experimentation and design.

By January 1918, the company employed 11,500 workers (Trimble, 2013, p. 183) and between July 1917 and March 1919 produced 5,221 aircraft and about

5,000 engines, most being sold to the Army. At its peak, more than 18,000 worked for the company (Trimble, 2013, p. 186). Despite the very large-scale production achieved by the Curtiss Company during the war, it did not last long once the war had ended. Peace in 1919 brought about a cancellation of wartime contracts, and in September 1920, the Curtiss Aeroplane and Motor Company underwent a financial reorganisation after it had been forced into receivership. Glenn H. Curtiss himself took the opportunity to cash out his stock in the company for US$32 million and retired from the industry to Florida. Curtiss remained as a company director, but in practice only acted as an adviser on design. Clement M. Keys gained control of the company, which later became the nucleus of a large group of aviation companies, eventually merging with Wright in 1929. Curtiss went on to found 18 corporations in Florida, as well as serving on civic commissions and donating extensive land and water rights. Also, he co-developed the city of Hialeah with James Bright and developed the cities of Opa-locka and Miami Springs, dying in 1930, in Buffalo, New York, of complications that arose from an appendectomy.

It's not entirely clear why Curtiss got out of the industry in 1920. Perhaps he understood that the next few years for the industry would be difficult ones and that it was most prudent to take what money he could and get out. Or perhaps he was tired with the large-scale production that had occurred during the war, with him being more the type of person who was in aviation for the excitement of its early development. Whatever the reasons, he avoided the traumas of the 1920s, which other pioneers of aviation struggled to overcome, looking elsewhere to exercise his entrepreneurial talents.

Curtiss' expertise lay in engine manufacturing, which he then applied to a key problem facing aircraft at the time – how to get a heavier-than-air plane to deliver sustainable flight. His competitive interests aligned with the speed that planes could bring and the money that could be won. Barriers encountered were overcome through network connections – partnering with Bell to fund early engine development and partnering with the Navy to deliver planes. While Curtiss' company struggled during the post-war dip before becoming part of an amalgamated group, Curtiss successfully cashed out his stock to pursue other entrepreneurial opportunities. It's notable that Curtiss concentrated more on development rather than production after the 1916 reorganisation, and even then realised that his own interest lay in this field, rather than in the running of a large-scale industrial concern.

The Wright brothers and Curtiss both came into aviation from running their own bicycle businesses, giving them experience in start-up businesses. In contrast to the Wright brothers, who worked within their own companies, Curtiss worked for a number of others (notably Alexander Graham Bell) before starting his own aviation business. Both the Wrights and Curtiss made extensive use of air shows, air competitions and their own flight schools to build connections and secure funds. While they dominated the early industry, they both struggled and failed to maintain control of their businesses after World War I, selling out by the early 1920s.

The Wright brothers and Curtiss battled for control over the early aviation market in America through aircraft patents and designs – both in court and out. By 1917, they had effectively blocked the building of the aircraft, a situation overcome by the United States Government's intervention through the forced creation of a patent pool. Like the Wright brothers, Curtiss left the industry before the start of the post-World War I contraction period of the interwar years to pursue other entrepreneurial interests.

4.4. Allan Haines Loughhead (Lockheed)

By the early 1920s, the Wright brothers and Curtiss had left the industry, but others amongst the pioneer aviators tried to stay on and make some sort of headway in the industry. One such individual was Allan Haines Lockheed (1889–1969). Lockheed's original name was Loughead, and he was born in Niles, California, in 1889, the youngest son of Flora and John Loughead. He had a sister, Hope, a half-brother, Victor, and a full brother, Malcolm. His mother, Flora Haines Loughead, was a well-known journalist and novelist, and after separating from her first husband, she took her children to Santa Barbara, California, where her sons became interested in flight through their experimenting with kites. Later, Flora moved the family to a fruit ranch near Alma, California, where the brothers became interested in the gliding experiments of Professor John J. Montgomery. The brothers only attended elementary school, but like the Wrights and Curtiss, they were mechanically inclined from an early age, but unlike the others, they were exposed to literature by their mother.

Allan's half-brother, Victor, became interested in motorcars and aircraft, and moved to Chicago where he became associated with James E. Plew, a car dealer. There, in 1909, he wrote a book, *Vehicles of the Air*, which became a popular treatise on aircraft design and aviation history. Allan's other brother, Malcolm, became a mechanic with the White Steam Car Company in San Francisco in 1904, and later in 1906, Allan also moved to San Francisco where he also became a mechanic and by 1909 was driving racing cars.

In Chicago, Victor convinced Plew to acquire the rights to one of the Montgomery gliders and to buy a Curtiss pusher biplane. Victor then convinced Plew to hire Allan who moved to Chicago and began to convert the Montgomery glider to a powered aircraft; subsequently a two-cylinder, 12 h.p. motor was installed on the glider. Allan's first flight took place in Chicago in 1910. After Plew lost his interest, Allan became a flight instructor with the International Aeroplane Manufacturing Company in Chicago and put on aerial exhibitions for 25 per cent of the gate receipts. The three brothers, Victor, Malcolm and Allan, worked together for a time on these ventures in Chicago but later they all moved back to California.

Allan Loughead returned to San Francisco in 1912 and went back to working again as an auto mechanic. There, he and Malcolm spent their spare time building a three-place seaplane to operate from San Francisco Bay. As they constantly ran out of money building the plane, they managed to convince Max Mamlock of the Alco Cab Company to invest US$4,000 in its construction, and, finally, after 18 months, in 1913, their Model G was christened the ALCO NO. 1. Allan Loughead flew it for the first time from the waters of the Golden Gate Bridge entrance to San Francisco Bay on 15 June 1913.

With the financial aid of Paul Meyer, Allan and Malcolm Loughead bought the Model G back in 1915 and opened a flying concession at the Panama-Pacific International Exposition in San Francisco. In early 1916, the Loughead brothers moved to Santa Barbara, where they were swamped by people who wanted to make their first flight. In addition, they undertook charter flights to the offshore islands, and local movie companies used the plane to take aerial footage. In 1916, the brothers also organised the Loughead Aircraft Manufacturing Company in Santa Barbara to build a huge 10-place, twin-engine F-1 flying boat for their aerial joy riding and sightseeing business (Boyne, 1999).

After the United States entered World War I in 1917, Allan travelled to Washington, DC, to try to get a Navy contract to build the F-1. The Navy told him that it would only purchase previously approved designs, but he did manage to get a contract to build two Curtiss flying boats and an agreement for the Navy to test the F-1. After the war was over in 1919, the Loughead Manufacturing Company Aircraft entered the small aircraft market with the production of a single seat, S-1 Sport Biplane. This aircraft was intended to be 'the poor man's airplane', and it featured an innovative moulded plywood monocoque fuselage for which the Lougheads, Northrop and Tony Stadlman received a patent. This aircraft was to make hundreds of flights, and it proved to be a successful technical design; however, although at an aircraft show in San Francisco thousands admired the plane, no one ordered any. In the post-war slump, the Lougheads found little business, and eventually the discouraged Malcolm took a job in the motorcar industry making brakes. Later, Allan also came to realise that the government's sale of war surplus aircraft had effectively killed the market for the production and sale of new aircraft and as a result, Loughead Aircraft was closed in 1920, and in the following years, its assets were sold.

For the next few years, Allan Loughead worked as the Los Angeles sales manager for Lockheed brakes, a company owned by Malcolm. Later, in 1922, he became a real estate salesman in the Hollywood area. Gradually, however, as the aircraft industry began to recover in the mid-1920s, Allan Loughead began to be drawn back in. In 1926, he returned in partnership with Jack Northrop when they decided to build a high-speed monoplane with a capacity to carry four passengers and a pilot in a streamlined fuselage using their patented monocoque construction. Loughead and Northrop set out to form an aircraft company and Loughead's accountant friend, Kenneth Jay, introduced them to Fred S. Keeler, a successful brick and ceramics manufacturer. After reviewing their proposal, Keeler agreed to help finance the project, and as a result, using US$22,500 from Keeler and US$2,500 from Loughead, the four formed the Lockheed Aircraft Corporation in December 1926, with Keeler as president, Loughead as vice president and general manager, Northrop as chief engineer and Stadlman as factory superintendent. They used the 'Lockheed' spelling in order to associate themselves with Malcolm Loughead's successful brake company, which also was using the altered name.

The company began its operations in January 1927 in a garage in Hollywood, and the first product of the company was the building of a successful high-speed, streamlined monoplane with an effective flight range of 1,000 miles, a cruising speed of 185 miles per hour, and capacity to carry six people (Hewson, 2013). This plane was known as the Lockheed Vega, and it was to remain the primary product of the Lockheed Corporation. It was manufactured in four-passenger and six-passenger variants and by April 1929, the company was producing five planes per week with a workforce of fewer than 300 employees. The retail sales price of these planes averaged about US$17,000 each.

The success of the design meant that a steady stream of orders for the new aircraft came, which necessitated a move, in March 1928, to new and larger facilities in Burbank, California (Hewson, 2013). Lockheed at this time received an order for 20 Vegas worth US$250,000, the largest commercial aircraft order to date, with the nation's fledgling airlines recognising the potential of the plane as an airmail and passenger plane. At the same time, Northrop designed the parasol wing, Air Express for Western Air Express, as an airmail and a passenger plane. The Vega and Air Express, and a variety of variants that stemmed from them, came to be used by some of the biggest names in aviation at the time including Art Goebel, Bob Cantwell, Frank Hawks, Amelia Earhart, Wiley Post, Roscoe Turner and Jimmy Doolittle. The sales of these aircraft meant that in 1928 company sales exceeded US$1 million.

In mid-1928, Jack Northrop left Lockheed Aircraft to start his own company, and Gerald Vultee (later to found Vultee Aircraft) became the Lockheed chief engineer. Part of the reason Northrop left the company was that Lockheed's management refused to invest in developing new metal aircraft and chose instead to try to maximise its revenue from the sale of its proven wood designs. Meanwhile, the Detroit Aircraft Corporation, a holding company with assets of US$28 million, began acquiring a portfolio of aviation companies and turned its attention to the Lockheed Company. In July 1929, Fred E. Keeler, who owned 51 per cent of Lockheed stock, decided to sell his holdings to Detroit Aircraft, the acquisition occurring through an exchange of stock. Unhappy with this turn of events, Allan Loughead resigned as president and general manager on 3 June 1929 and later sold his Detroit Aircraft stock for US$23 a share. This proved to be a stroke of luck for Lockheed, because with the stock market crash in October 1929, the Detroit Aircraft stock fell to 12.5 cents a share, and with the collapse in demand for aircraft, by 1932, Lockheed Aircraft was bankrupt.

A group of investors headed by brothers Robert Gross and Courtlandt S. Gross, with the addition of Walter Varney, bought the Lockheed Company out of receivership in 1932, and Allan Lockheed returned to the company as a consultant, but with no formal management role in his namesake company. Instead in 1930, he formed the Lockheed Brothers Aircraft Corporation in Glendale, California, and developed the experimental Olympia Duo-Four, a five-place, high-wing monoplane with two engines mounted side by side in the wood monocoque fuselage nose. This plane had a plywood-covered wing and wheel pants and the fuselage was covered with a two-piece moulded-under-pressure plywood skin. Allan

Lockheed made a number of flights in this plane, demonstrating its extremely safe single-engine performance. This fourth commercial aircraft business founded by Loughead survived until 1934, and in the same year, Loughead legally changed his name to Allan Lockheed.

In 1937, Lockheed yet again formed another company, the Alcor Aircraft Corporation in San Francisco, and developed the Alcor C-6–1, an eight-place, low-wing plane that also had excellent single-engine performance. Unfortunately, the prototype C-6–1 was lost over San Francisco Bay and the Alcor Company eventually folded in 1939.

In 1941, Lockheed became vice president of the Berkey & Gay Furniture Company in Grand Rapids, Michigan, where he served as general manager of the Aviation Division and Director of Aircraft Engineering. In October 1942, Lockheed became the general manager of the Aircraft Division of Grand Rapids Store Equipment Company, which made parts for Navy fighters. After the war was over, Allan Lockheed continued his career as a real estate salesman in California, while also occasionally serving as an aviation consultant. In the mid-1950s, the Lockheed Aircraft Corporation asked Allan Lockheed to return as a consultant, and in 1961, Lockheed moved to Tucson, Arizona, where he lived in semi-retirement; however, he continued to do consulting work with the Lockheed Aircraft Corporation. Allan Lockheed died of liver cancer in Tucson in 1969 at the age of 80.

Although Allan Lockheed stayed in the industry considerably longer than the Wrights and Curtiss, he experienced considerable difficulty in doing so. Over the years he formed and disbanded a range of companies, showing an interest (and ability) far more for the design of new aircraft rather than the running of substantial manufacturing enterprises. He did play an additional important role in the industry by working with people such as Vultee and Northrop, who both went onto make important contributions to the development of the industry.

Allan Lockheed benefitted from the patent pool created by the United States Government and from the exit of the Wright brothers and Curtiss from the market, but the dominance of the legacy designs from these inventors continued to block the introduction of new design alternatives. Lockheed connected early with the public by setting up and offering aerial sightseeing for the public and movie producers. This was one of many changes in jobs which saw Allan Lockheed working for others at the start and end of his career, while working for his co-owned businesses in the central part of his career.

The shift in Lockheed's fortune occurred as a result of the development of the high-speed, long-distance travelling monoplane, which the fledgling American commercial flying industry quickly realised could fill a gap in their design portfolio, effectively opening up a new area of the industry which he continued to shape through the remainder of his career. Unlike the Wrights and Curtiss, the context in which Lockheed operated was one of incumbent-dominant designs and limited resources to stimulate new design. Despite this, his entrepreneurial vision allowed him, together with his partners, to pursue and develop the emerging commercial airline industry by cross-pollinating ideas from one part of a shrinking military aviation industry to an emergent commercial aviation market. Lockheed's extensive networks allowed him to access the resources he needed during difficult times.

4.5 Glenn Luther Martin

One pioneer whose long-term business career was rather more successful than those mentioned so far was that of Glenn Luther Martin (1886–1955). Martin was one of the most successful aircraft pioneers, retaining control of his company for 40 years. Glenn L. Martin was born in Macksburg, Iowa, on 17 January 1886 to Clarence and Minta Martin. At the age of two, Martin's family moved to Salina, Kansas, so that his father could run a wheat farm, and by the age of six, he had become interested in flying kites, turning his mother's kitchen into a 'factory' to produce them. Martin also began using sails on everything from ice skates to wagons, and even his bicycle, to move faster with less effort.

He attended and studied business at Kansas Wesleyan in Salina, Kansas, and as he grew up, his fascination for flight intensified. In 1909, he decided to build an aircraft himself based on the Curtiss June Bug, but it was destroyed on its first test flight. For his next effort, Martin used silk and bamboo in the aircraft's construction and this aircraft made a short flight. Martin was often assisted by his mother, who held a lamp over him in the evenings as he worked on his early aircraft.

On 10 May 1912, Martin flew a self-built seaplane from Newport Bay, California, to Avalon on Santa Catalina Island, then back across the channel. This broke the earlier English Channel record for over-water flight, a distance of 68 miles. He picked up a bag of mail on the island on the way and was presented with a US$100 prize for his achievement. In 1913, Martin was not as fortunate while competing in the Great Lakes Reliability Cruise, a 900-mile race of seaplanes around the Great Lakes. Martin's pontoon hit a wave at high speed and low altitude, causing the plane to somersault and sink to the bottom with Martin, who escaped and attempted to salvage the plane to finish the race.

Using the prize money that he had won, as well as his reputation as a flyer, he founded the Glenn L. Martin Company in 1912, and in the same year, he built an aircraft factory in an old Methodist church in Los Angeles, California. Lawrence Bell, who later formed Bell Aviation, was appointed as shop foreman (Glenn Martin, 1960). To make money to finance this business, he began stunt-flying at fairs and local airfields and also got involved in the early movie industry. This he did after seeing an advertisement for a pilot/aircraft owner to play a role in a movie. Sensing an opportunity to market his aircraft to a larger audience, he replied to the ad and got the part, playing the role of a dashing hero in the movie *A Girl of Yesterday* (1915) starring Mary Pickford. In addition to flying Pickford around in his aircraft, he had a scene where he had to kiss Frances Marion, who later became a legendary Hollywood screenwriter (see Figure 4.3).

While he was working in the movies, his company began designing aircraft for the military in 1914, first building trainers for the United States Signal Corps. An early success came during World War I with the production of the MB-1 bomber,

Figure 4.3 Douglas Gerrad, Glenn L. Martin and Mary Pickford

with the MB-2 and others also subsequently being successful. In 1916, he merged his company with the original Wright Company, forming the Wright-Martin Aircraft Company; however, his time with the new company was not happy, and so he left in 1917 to found a second company, Glenn L. Martin Company. The Wright-Martin Company, once it was formed, left Martin as the vice president in charge of production, but the company made the decision to confine production to making engines, which convinced Martin to leave the company. A group of Cleveland investors raised money for him to reform his company, and Donald Douglas (who later founded the Douglas Aircraft Company) worked for the him, along with Bell as plant manager and J.A. Kindleberger as the chief designer. After the war ended Martin brought out the Cleveland financiers, and small orders from the post office and Navy kept him going in business until June 1920. The Army then ordered some MB-2s and small Navy orders kept the company going from 1922 to 1924. The company then built more bombers and by 1928 its workforce had risen to around 1,000.

In 1925, Martin began to think about looking for a larger site to enable expansion, and to this end, a site in Middle River, Maryland, was chosen. It subsequently took three years to acquire the land needed from 45 property owners, which involved convincing the owners that this was going to bring a booming industry to the region that would provide many jobs for the area. In 1928, the Glenn L. Martin Company finally made the move to Maryland. The company continued to concentrate on the building of bombers through the 1930s and Martin stayed in control of the company through World War II, managing his company well into the post-war world. After World War II, the company struggled to make sales and ended up borrowing from the United States Government to survive. Eventually, the company was refinanced, but with Martin no longer as president (Still, 1964, p. 236). The basic problem at this time for the company was that it struggled to improve on the designs for bombers in the post-World War II era. Competition for military contracts was intense at this time, and a lack of design success usually meant the failure of a firm.

Despite the company's problems, Martin continued his involvement in the industry until his death, which arose from complications of a stroke on 5 December 1955 in Baltimore, Maryland. Throughout the latter part of his life, he maintained the independence of his company, perhaps mindful of the unhappiness he had experienced with the Wright-Martin merger of 1916. After the death of its founder, the company eventually merged with the American-Marietta Corporation in 1961, which became the Martin Marietta Corporation. This company, in turn, merged with the Lockheed Corporation in 1995, forming Lockheed Martin, a major United States aerospace and defence contractor today.

Martin was fortunate in that he was both a flyer and a designer, winning awards and competitions for both. This meant he was able to fly and promote his planes through stunt-flying and film (similarly to Allan Lockheed), and to be involved in manufacturing bombers for the military for both World Wars. These skills enabled him to establish a niche in the industry and thereby to deliver jobs, an airport and a booming industry sector at the time of the Great Depression. He was similar to the other pioneers in that he was inventive and mechanically minded from a very early age and enjoyed hobbies that involved some degree of speed. He differed in that he not only completed high school but also went on to college, although there he received a business education rather than an engineering one as a number of later pioneers like Douglas and McDonnell did. In that sense, he was a linking personality between the very first entrepreneurs with their limited education and technical hobbies and experiences and the later ones that came to prominence in the industry in the 1920s and 1930s with their engineering degrees.

4.6 Conclusion

In the United States, the original pioneers of flight like the Wright brothers, Martin, Curtiss and Loughead had similar backgrounds in that they all enjoyed adventurous hobbies, had experience at tinkering or working with bicycles, motorbikes and motorcars, and all tended to work in small businesses of their own rather than pursue conventional engineering careers. At a time when many people were reluctant to leave established jobs and careers, most of the pioneers seemed very adventurous in terms of their inclination to switch and change jobs. Their education also tended to be scant (with perhaps the exception of Martin), but their early work experience exposed them to many of the new technical developments taking place at this time.

Although the pioneers of aviation in the United States aspired to create sustainable commercial businesses that built aircraft and allowed them to pursue their aviation interests, each found it challenging. The pre-World War I era was a difficult one for the pioneers of aviation who wanted to use their knowledge in entrepreneurial pursuits. With the exception of the Wright brothers, entrepreneurs like Curtiss, Martin and Loughead struggled to attract the necessary financing to establish viable businesses. Private sums were the source of much of the original capital in the industry, as well as money earnt from other small businesses, air exhibitions, prize money and the revenue generated from flying schools.

World War I for the United States was a relatively short affair and probably only Curtiss was able to take full advantage of the demand it created. He had the foresight to cash out at the end of the war, probably because he had an inkling that the exciting, pioneering phase of the industry's development was over and that the wartime mass-scale industry was the future – one that he wasn't all that happy to be involved in. Orville Wright also found it difficult to run his business after the death of his brother and conflict with his sister, and sold out of it in 1915. Loughead, in contrast, retained his interest in the industry, despite a series of setbacks, illustrating that the industry itself had a particular fascination for a number of the pioneers. It's interesting to see, therefore, that a number of the entrepreneurs proved unable to make the transition from innovator or designer to manager of large-scale manufacturing.

The market for aircraft prior to World War I was tiny, and America's involvement in the war was too short to change the situation much. What it did do, however, was create a substantial surplus of aircraft in the immediate post-war years, which damaged the surviving companies. United States military sales in the 1920s were meagre and although they helped to keep a few small companies in the industry alive, these still struggled. During the war itself, airframes were built by a variety of companies, mainly from the motorcar industry. In 1917, military production had grown from a total of 400 in 1916 to over 2,000 aircraft. In 1918, total production was 14,000, then slashed in 1919 to fewer than 800. After the war was over military procurement was greatly reduced and by mid-1924, the United States Army Air Service possessed only 74 pursuit (fighter) aircraft. The resulting decline in military demand had a devastating effect on the industry. Production of aero engines also fell and by 1922 production had fallen to 263 (Rae, 1968, p. 3). The dramatic decline meant that military procurement in the 1920s tended to be very small scale, with specific types of aircraft making more economical (and potentially more profitable) longer production runs impossible. Through the 1920s, there were only a few companies operating in the United States, with the Curtiss Company surviving by producing a large share of the fighters for the military, while Glenn L. Martin dominated the bomber market and Wright mainly manufactured aero engines.

Most of the aviation manufacturers, therefore, struggled through the 1920s, making a small number of recreation aircraft and aircraft for the military. This meant that the industry up until the late 1920s in many ways lagged behind that of France and the United Kingdom, where military purchases of aircraft gave the industry more of a boost. For that reason, before going on to look at the new breed of innovators and entrepreneurs who entered the American industry in the 1920s and 1930s, it is important to see how developments were taking place in Western Europe at this time. In both France and the United Kingdom, aviation pioneers who started up aviation manufacturing firms prior to World War I continued on in the industry in the 1920s and 1930s.

References

Anderson, J. D. (2004). *Inventing Flight: The Wright Brothers and their Predecessors*. Baltimore, MD: Johns Hopkins University Press.

Ash, R. (1974). *The Wright Brothers*. London: Wayland.

Boyne, W. J. (1999). *Beyond the Horizons: The Lockheed Story*. New York, NY: St Martin's Griffin.

Casey, L. S. (1981). *Curtiss: The Hammondsport Era 1907–1915*. New York, NY: Crown Publishers.

Crouch, T. D. (2003). *The Bishop's Boys: A Life of Wilbur and Orville Wright*. New York, NY: W. W. Norton & Company.

Glenn, L. Martin Co. (1960). *Box Kites to Bombers: The Story of the Glenn L. Martin Company*. Baltimore, MD: Glenn L. Martin Co.

Gollin, A. M. (1984). *No Longer an Island: Britain and the Wright Brothers, 1902-1909*. Palo Alto, CA: Stanford University Press.

Harley, R. W. (1967). *Glenn L. Martin: Boy Conqueror of the Air*. Indianapolis, IN: Bobbs-Merrill.

Hatch, A. (2007). *Glenn Curtiss: Pioneer of Aviation*. Guilford, CT: Lyons Press.

Howard, F. (1988). *Wilbur and Orville: A Biography of the Wright Brothers*. New York, NY: Ballantine Books.

McPherson, S. S. and Gardner, J. S. (2004). *Wilbur & Orville Wright: Taking flight*. Minneapolis, MN: Carolrhoda, Inc.

Mitchell, C. R. and House, K. W. (2001). *Glenn H. Curtiss: Aviation Pioneer*. Charleston, SC: Arcadia Publishing.

Rae J.B. (1968). *Climb to Greatness: The American Aircraft Industry 1920-1960*. Cambridge, MA: MIT.

Roach, E. J. (2014). *The Wright Company: From Invention to Industry*. Athens, OH: Ohio University Press.

Roseberry, C. R. (1972). *Glenn Curtiss: Pioneer of Flight*. New York, NY and Garden City: Doubleday & Company.

Shulman, S. (2002). *Unlocking the Sky: Glenn Hammond Curtiss and the Race to Invent the Airplane*. New York, NY: HarperCollins.

Still, H. (1964). *To Ride the Wind: A Biography of Glenn L. Martin*. New York, NY: Julian Messner.

Tobin, J. (2004). *To Conquer the Air: The Wright Brothers and the Great Race for Flight*. New York, NY: Simon & Schuster.

Trimble, W. F. (2013). *Hero of the Air: Glenn Curtiss and the Birth of Naval Aviation*. Annapolis, MD: Naval Institute Press.

Vander Meulen, J. A. (1991). *The Politics of Aircraft: Building and American Military Industry*. Lawrence, KA: University of Kansas Press.

5 The French connection

Louis Blériot, Raymond Saulnier, the Farman brothers and Marcel Dassault

5.1 Introduction

Although powered, heavier-than-air flight originated in the United States, it was the European powers that embraced flight most readily in the early years. Well before the outbreak of World War I a number of European countries moved ahead of the United States in developing leading-edge technologies. In France, for instance, some of the pioneer flyers included such men as Alberto Santos-Dumont, Léon Levavasseur, Gabriel and Charles Voisin, Henri Farman (see Figure 5.1), Ambroise Goupy, Louis Blériot, Robert Esnault-Pelterie, Lois Breguet, Edouard de Nieport and Louis Bechereau. These men all began to manufacture aircraft in the period leading up to the outbreak of World War I.

Prior to World War I, France was undoubtedly the leader in the development of the industry. To begin with, it was the country that had the first designers that were to establish independent firms dedicated to building aircraft. The earliest company of this type was probably Gabriel Voisin and Ernest Archeadcon's Syndicat d'Aviation, which was formed as early as 1905. Blériot joined Voisin and formed Blériot-Voisin later in the same year, but the two parted company in 1906, with Voisin buying out Blériot's shares in the company. In November 1906, Gabiel Voisin and his brother followed on by forming Voisin Freres, the first commercial aircraft manufacturing company. The firm built gliders and aircraft and built around 20 aircraft before the outbreak of World War I. Blériot, in turn, formed his own company and built a series of popular aircraft around this time (Bakewell, 2010; Gibbs-Smith, 1953).

Although private buyers existed in France the French Army, more than any other military service in Europe, recognised the potential uses of aircraft. By 1912, it had increased its military inventory to over 250 planes, which it used for military reconnaissance. These purchases helped to keep a number of the small French companies going, and by 1914, 20 small firms had entered the industry, making France the world leader in aircraft production (Chapman, 1990, p. 18). During the war, aviation became a major part of the French military and by 1918 the industry employed 200,000 workers and built more than

Figure 5.1 Henri Farman after winning the Deutsch Prize 1911

Source: © Library of Congress. Bain News Service, Publisher. Henri Farman after winning Deutsch prize of aviation, ND Photo/ND Photo, 1908. Date created or published later by Bain. Photograph. www.loc.gov/item/2014680044/

51,000 planes, and 92,000 aero engines. This made the French industry the world's largest, and at the end of the war in November 1918, France had the world's largest air force.

Because of the early dominance of the French industry, the purpose of this chapter is to look at the involvement of the French aviation pioneers in the creation, or what was to become the world's leading-edge industry. It also looks at the way in which these enterprises gradually lost their competitive edge and the French pioneers gradually faded from prominence in the industry.

5.2 The French industry

Despite having the world's largest air force during World War I, in the immediate post-war years, the French industry suffered, much as the American one had done. The surplus of aircraft left over from the war meant that the French aviation workforce shrank from 180,000 in 1918, to 100,000 in 1919, to only 5,200 in 1920 and to a minuscule 3,900 by 1921 (Truelle, 1969; Chapman, 1990, p. 19). At the same time, the French Air Force shrank in size to employing 39,055 personnel, maintaining 3,940 aircraft by the end of 1920. Despite this fallout in the industry, a few French firms survived and France's leading role in the aviation industry survived into the late 1920s, with several French firms (Potez, Breguet, Hispano-Suiza and Gnome-et Rhône) remaining major competitors in world markets.

A variety of strategies were undertaken by the French aircraft manufacturing companies to survive in the years immediately after World War I. Many of the firms that had entered the industry during the war to take advantage of wartime demand returned to their pre-war production of other products. Other firms survived by investing their wartime profits and making other products. Other firms still took part in mergers at this time, the most notable being that of Louis Blériot's with SPAD in 1920 (Barros, 2011; Chadeau, 1988). In the late 1920s, however, France still had at least 23 firms making airframes and about ten making aero engines. Most of these firms depended on military orders to survive and the French Government spread its relatively small orders around them to keep the industry competitive and alive. This approach allowed the French Government to maintain capacity for any potential expansion, while also avoiding becoming too dependent on a few suppliers. The government, however, failed to expand the market sufficiently by either building a strong air force or by stimulating growth in the airline business, but instead simply acted to help the industry to survive in a semi-stagnant condition.

In the case of the French airlines by the late 1920s, there were five major airlines, each heavily subsidised by the state, all of which faced chronic financial difficulties, as well as stiff foreign competition. These airlines provided some alternative demand for aircraft to that of the military but on too small a scale to make much difference.

Several of the pre-war pioneers such as Louis Breguet, the Farman brothers and Fernand Loire still ran firms in the French aviation industry at this time. During the war, however, a second generation of builders entered the industry; men like Henry Potez, and Marcel Bloch. These men both graduated from the new École Superieure de l'Aeronautique in Paris, joined the army engineering corps and used their positions to launch firms of their own. Émile Dewoitine, a brilliant young engineer, had a similar start in Toulouse, during the war rising to be a technical director for Pierre-Georges Latécoère's company before starting his own firm. Breguet, Potez, Farman and Liore et Olivier remained firms that were competitive internationally, but many firms only just survived.

In the domestic market, the widest possible distribution of the French Government's aeronautical budget was used to keep as many firms as possible operating. This was justified as creating competition, innovation and the maintenance of capacity, although it was probably also due to political pressure to maintain employment in a variety of centres. This was notably a similar strategy to that undertaken by the United States Government in the 1920s. The other feature of the French Government's expenditure on aviation was that relatively little was invested in research, concentrated instead on buying proven aircraft types (Barros, 2011, p. 93).

By 1927, employment in the industry had risen to 11,000, mainly supported by military purchases. At the same time, in contrast, the German industry, which had been prohibited from building military aircraft by the Treaty of Versailles, was developing an advanced commercial airlines network, to support its aviation industry. By 1928, Germany boasted a network of domestic air routes covering 60,000 kilometres, compared to France's 3,000 kilometres. Semi-government airlines in the United Kingdom and Germany were also threatening to squeeze out French airlines from international air routes.

Consolidation of the industry was difficult as most companies had built up capacity during the war, and many leaders of industry had strong personal commitments to aviation and their own organisations. Like in the United States and the United Kingdom, mergers tended not to occur until the founders of the companies had retired. Two mergers that did, however, occur later in the 1920s led to the creation of Loire-Nieuport and Potez-CAMs. Another notable restructuring took place in 1932 when the company Morane-Saulnier failed.

Despite the wartime edge, by 1925 the French industry had begun to lag behind the British and American manufacturers and by 1935, they were using all metal monocoque structures. In the United States and the United Kingdom, reliable air-coiled radial engines and in-line engines came onto the market together with new cowlings and variable pitch propellers. These developments were supported by sophisticated test facilities, including the use of wind tunnels.

In the French case, these technological advancements were eventually adopted – in 1931, French firms made the breakthrough to all metal airframes and by 1933 some prototypes were able to reach a speed of 350 kph. However, by the early 1930s, French aircraft manufacturers were losing ground to American firms, which were better equipped for aggressive marketing, corporate mergers and mass production.

The political climate in France also was more unstable, and in 1936, the French aviation industry was nationalised as a result of political and industrial controversy surrounding the role of the armaments industry throughout the 1930s. This was despite the French industry having a relatively quiet industrial record and the overall more moderate nature of the unions in this industry. There was, however, a fairly wide degree of support in France for this measure with Daladier's nationalisation bill passing through parliament by a substantial majority. The legislation meant that about 80 per cent of the airframe sector was reorganised into five companies, each two-thirds owned by the government (Chapman, 1990; Higham, 2012; Baughen, 2018). Younger managers such as Henry Potez and Marcel Bloch were given leading positions in the nationalised companies, but several of the older pioneers of the industry including Louis Breguet, Robert Morane and Raymond Saulnier were passed over as directors of the nationalised sector. In contrast to the airframes section of the industry, the two major engine manufacturers were left in private hands, the reason being their strong affiliations to the motorcar industry. In the lead up to the outbreak of World War II, a number of the French aviation pioneers continued to play a key role despite all of these changes and industrial development at a time when rearmament was slowly starting to pick up (see Figure 5.2).

Figure 5.2 Assembly line in Bourges, France (to the right: Potez 631)

Source: Picture from the book *Curtiss Hawk 75* by Jean Cuny and Gérard Beauchamp. Docavia/Editions Larivière

5.3 Louis Charles Joseph Blériot

One of France's best-known pioneer aviators was Louis Blériot (1872–1936). Born in Cambrai, Blériot was the first of five children born to Clémence and Charles Blériot. In 1882, at the age of 10, Blériot was sent as a boarder to school at the Institut Notre Dame in Cambrai, where he was successful at winning class prizes, including one for engineering drawing. At the age of 15, he went to the Lycéeat Amiens, and after passing the exams for his baccalaureate in science and German, he tried to enter the prestigious École Centrale in Paris. One of France's preeminent colleges, its founding vision was to promote the development of 'industrial science, and the practical application of recent major scientific discoveries'. It also aimed to educate multidisciplinary engineers and prepare them for leadership positions in the emerging industrial sector. This meant that his education was a very suitable one for preparing him for his future work in aviation. Entrance to the institute was via a very demanding exam for which special tuition was necessary, and so Blériot spent a year at the Collège Sainte-Barbe in Paris in preparation. He passed the exam, doing especially well in the tests of engineering drawing ability. After three years of demanding study at the École Centrale, Blériot graduated and then began a term of compulsory military service, spending a year as a sub-lieutenant in the 24th Artillery Regiment, stationed in Tarbes in the Pyrenees.

After he completed his military service, he got employment with Baguès, an electrical engineering company in Paris, but later left the company after developing the world's first practical headlamp for motorcars using a compact integral acetylene generator. Blériot was a persistent inventor and used the invention of this product to start his own business. In 1897, therefore, Blériot opened a showroom for headlamps in Paris and the business was successful, and soon he was supplying his lamps to both Renault and Panhard-Levassor, two of the foremost motorcar manufacturers of the day. Later he went on to use much of the money that he made in this enterprise to finance his attempts to build a successful aircraft (Elliot, 2000; Lockwood Marsh, 1935).

Like so many of the early entrepreneurs of the aviation industry, Blériot had a strong technical education but was disinclined to follow a conventional career in engineering working for established companies. After his military service and a brief period working for the electrical engineering company, he set up his own firm.

Blériot had first become interested in aviation while he was at the École Centrale, but his serious experimentation with designing aircraft may have been sparked by seeing Clément Ader's Avion III at the 1900 Exposition Universelle. His first experiments with building aircraft were with a series of ornithopters (early helicopters), which were unsuccessful, and in April 1905, Blériot met Gabriel Voisin. Blériot was a spectator at Voisin's first trials of a floatplane glider he had built and flown in 1905, and the success of these trials prompted him to purchase a similar machine from Voisin. After some experiments with this aircraft, he suggested that Voisin should enter into partnership with him, which Voisin accepted. The two men established the Ateliers d'Aviation Edouard Surcouf, Blériot et Voisin, and in the following year, the company built two unsuccessful powered aircraft, the Blériot III and the Blériot IV, the latter being largely a rebuild of the former. Both of these aircraft were powered with the lightweight Antoinette engines being developed by Léon Levavasseur.

This partnership did not last long and after Voisin bought out Blériot's shares the latter went onto form his own company to build aircraft. The first of these aircraft, the Blériot V, was first tried on 21 March 1907 when Blériot limited his experiments to ground runs, which resulted in damage to the undercarriage. Two further ground trials, also damaging the aircraft, were undertaken, followed by another attempt on 5 April. The flight was only of around 6 metres, but more trials followed, and, eventually, on 19 April 1907, the aircraft left the ground. Blériot, unfortunately, crashed the plane while attempting to land it.

Further aircraft development occurred with the Blériot VI, a tandem wing design, first tested on 7 July, when the aircraft failed to lift off. Blériot then enlarged the wings slightly, and on 11 July a short successful flight of around 25 to 30 metres was made, reaching an altitude of around 2 metres. This was Blériot's first truly successful flight and further successful flights took place that month. After crashing this aircraft Blériot abandoned it and, concentrating on his next machine, Blériot continued working on a series of new aircraft, each an improvement on the previous one until he was able to produce Blériot XII, a high-wing two-seater monoplane which was first flown on 21 May 1909.

Blériot's attempts to fly were generally well regarded by the general public and on 16 June 1909 Blériot and Voisin were both given official recognition when they jointly were awarded the Prix Osiris, which was awarded by the Institut de France every three years to the Frenchman who had made the greatest contribution to science. Three days later, on 19 July, he informed the *Daily Mail* of London of his intention to make an attempt to win the thousand-pound prize offered by the newspaper for a successful crossing of the English Channel in a heavier-than-air aircraft. The *Daily Mail* prize was first announced in October 1908, with a prize of £500 being offered for a flight made before the end of the year. When 1908 passed with no serious attempt being made to claim the prize, it was doubled to £1,000 and the offer extended to the end of 1909. On 25 July 1909, he made his crossing, the flight taking 36 minutes and 30 seconds.

This flight in 1909 made Blériot world famous for making the first flight across the English Channel which in turn brought about an immediate transformation of the status of his company, the Recherches Aéronautiques Louis Blériot. By the time of the channel flight, Blériot had invested a considerable amount of his own money on his aviation experiments, the cross-channel flight allowed him to recoup some of this investment due to a steady stream of orders coming in. By the end of 1909, orders for over 100 aircraft had been received by his company and between 1909 and the outbreak of World War I in 1914, Blériot produced about 900 aircraft, most of them variations of the Type XI model, making his company the world's largest producer of aircraft in the pre-World War I period. Blériot monoplanes along with Voisin-type biplanes, with the latter's Farman derivatives, dominated the pre-war aviation market in France.

Blériot, concentrated on the production of monoplanes, unusual at this time as there were concerns about the safety of them in general, not just in France but also in the United States and the United Kingdom. At one point the French Government even went as far as to ground all monoplanes in the French Army from February 1912 after accidents to four Bleriots, but, subsequently, it lifted this ban after trials in supported Blériot's analysis of the problem and led to a strengthening of the aircraft's landing wires. The brief but influential ban on the use of monoplanes by the Army (though not by the Navy) in the United Kingdom was triggered by accidents to other manufacturer's aircraft; Blériot's were not involved.

Another problem faced by Blériot was that of the patents issue. In the pre-World War I period, the legal disputes that impeded the development of the industry in the United States began to flow into France as well. Along with five other European aircraft builders, from 1910, Blériot was involved in a five-year legal struggle with the Wright brothers over the latter's wing-warping patents. The Wrights attempted to have their intellectual property rights, which had been established in the United States, extended into European jurisdictions as well. Eventually, however, the Wrights' claim was dismissed in the French and the German courts, which allowed the industry in France to develop unimpeded.

From 1913, Blériot's aviation activities were handled by his company, Blériot Aéronautique, based at Suresnes, which continued to design and produce aircraft up until the nationalisation of most of the French aircraft industry in 1937, when it was absorbed into Société Nationale de Constructions Aéronautiques du Sud

Ouest (SNCASO). The success of the company prior to World War I meant that in 1913, a consortium led by Blériot was able to buy out the Société pour les Appareils Deperdussin, an aircraft manufacturer, and he became the president of the company in 1914. He subsequently renamed it the Société Pour L'Aviation et ses Dérivés (SPAD), and this company produced some of the most famous of the World War I fighter aircraft, such as the SPAD S.XIII. After the war ended, the Blériot and SPAD companies were merged, and in the 1920s and 1930s, the joint company produced a range of aircraft used by several European airlines.

Because of his widespread reputation amongst the general public outside of France after his channel crossing, Blériot attempted to spread his operations abroad. Before World War I, Blériot opened British flying schools at Brooklands, in Surrey and at Hendon Aerodrome. Realising that a British company would have more chance to sell his models to the British Government than a French one, in 1915, he set up the Blériot Manufacturing Aircraft Company Ltd. in England (Sanger, 2008). The hoped-for orders did not, however, follow, as the Blériot design was by then seen as outdated by the British and following an unresolved conflict over control of the company, it was wound up on 24 July 1916. Even before the closure of this company Blériot was planning a new venture in the United Kingdom. Initially named Blériot and SPAD Ltd. and based in Addlestone, it became the Air Navigation and Engineering Company (ANEC) in May 1918. ANEC survived in a difficult aviation climate until late 1926, producing Blériot-Whippet cars as well as several light aircraft.

After the war, Blériot remained active in the aviation business until his death on 1 August 1936 in Paris due to a heart attack. In the 1920s, however, he tended to fade somewhat as his company was gradually superseded by others. The 1920 restructuring of his company sidelined him somewhat; however, he still remained working in the industry despite the setbacks. Blériot was similar to a number of other aviation pioneers in that he had a strong engineering education and background, used money from private interests including the winning of prize money to finance his initial efforts and stayed in the industry as long as he could, despite the collapse of the industry after World War I ended.

Like other aviation entrepreneurs, Blériot's early involvement with exhibitions, flight competitions and trials of others' planes exposed and connected him to the technologies and innovators of the time. Like the Wright brothers, he explored opportunities outside his country to build his business and reputation. He had several businesses in related industries some of which continued even after he had moved into aircraft design and construction where he quickly established a dominant position. He was adept at keeping himself in the media eye by competing and winning their prizes, making him an obvious supplier choice for the French Government for World War I. His experimentation brought him into conflict over patents with the Wright brothers but being in Europe offered some protection from the Wrights' control. His British founded flying schools in key aviation areas in the United Kingdom secured him a profile in the United Kingdom, even if he was not able to capitalise on it through British military contracts.

5.4 Raymond Victor Gabriel Jules Saulnier

Another of the prominent French aviators who was successful at manufacturing aircraft in the early years of the industry was Raymond Saulnier (1881–1964). Saulnier was a graduate of the École Centrale Paris and first collaborated with Louis Blériot on the Blériot XI, which was used for the first channel crossing. In October 1911, he founded the Morane-Saulnier company in partnership with the Morane brothers, and it was in working for this company that he designed many aircraft for which he filed numerous patents (the Société Anonyme des Aéroplanes Morane-Saulnier). He also at this time designed and built the aircraft in which Roland Garros made the first crossing of the Mediterranean on 23 September 1913.

He took an interest in publicising the efforts of the industry and as such was the chief editor of an aviation periodical and wrote *Etude, centrage et classification des Aéroplanes*, which was considered an authoritative work at the time on the aircraft industry. During World War I, Saulnier came up with the idea of using a device to allow the synchronisation of the firing a machine gun through a propeller, around the same time that Fokker was working on a similar device in Germany.

After World War I, Morane-Saulnier produced a number of designs for training and general aviation, but with the threat of war in the late '30s, he once again turned to the production of military aircraft. During the late 1920s and early 1930s, his company produced a number of parasol wing fighters, including the M.S.230 and M.S.315, but all were of limited performance and were relegated to training duties. Moraine-Saulnier had much more success with its dramatically modernised M.S.406, which was the French Air Force's most numerous fighter plane at the start of the war; unfortunately, however, the 406 was advanced only at the time of its introduction in 1935 and suffered terribly against the more modern Messerschmitt Bf 109s which it faced in 1940.

Although flyers like Blériot were prominent in World War I, many such as him faded from the industry in the 1920s. Saulnier was an exception as he was able to keep his company competitive into the 1920s and 1930s. Like the other French companies, he struggled in the 1930s, but he remained independent of the nationalised industry after 1936 and managed to reconstitute his company after the war.

After the defeat of the French Army in 1940, Morane-Saulnier was operated under German control and it built a number of German types, including the Fieseler Storch, known after the war as the Morane-Saulnier MS.500 Criquet. Morane-Saulnier also produced a number of trainer and civilian aircraft models, the best known of which was the successful Rallye series of four-seat STOL semi-aerobatic tourers.

Saulnier continued to manage his company after the war and between 1945 and 1964; under his direction, Morane-Saulnier produced over 1,000 aircraft and some 30 prototypes, his career in aviation extending into the post-World War II period just as Glenn L. Martin's did in the United States, as did a number of British aviators. He went on to personally manage Morane-Saulnier until 1961, and in

1962, the company filed for bankruptcy before being integrated firstly into Sud-Aviation, of which it became a subsidiary, then in 1966, its civilian models were spun off to form the Société de Construction d'Avions de Tourisme et Affaires, which was eventually purchased by Aérospatiale.

Saulnier was both an active contributor in creating and disseminating emerging aviation knowledge as a designer, patent holder and editor of an aviation periodical. This put him in a prime position to keep a finger on the pulse of aviation development. Under his management and leadership his company survived the two world wars and the interwar periods, demonstrating his ability to negotiate the changing contextual environment and protecting his company from being nationalised.

5.5 Farman brothers

Another amongst the early pioneers were the Farman brothers, Henri, Richard and Maurice. Their family was British in origin and they took French nationality in the 1930s. Each was born in Paris, France, the sons of a well-to-do British newspaper correspondent working there and his French wife. They were all educated at home and Henri (1874–1958) originally trained as a painter at the École des Beaux Arts, but quickly become obsessed with the new mechanical inventions that were rapidly appearing at the end of the nineteenth century. Since his family had money, he was able to pursue this interest as an amateur sportsman, and in the 1890s, he became a championship cyclist, and at the turn of the century, he discovered motor racing, competing for Renault in the Gordon Bennett Cup. His love of activities involving speed seems a fairly regular occurrence amongst the early pioneers of aviation. When the Voisin brothers started their aircraft construction business in 1907, Henri Farman was one of their first customers, ordering a copy of the aircraft that had been built for Leon Delagrange. He used this aircraft, the Voisin 1907 biplane, to set numerous official records for both distance and duration (see Figure 5.1). These include being the first to fly a complete circuit of one kilometre (13 January 1908, winning the 50,000-franc Grand Prix d'Aviation offered by Henri Deutsch de la Meurthe) and first to make a two kilometres circle (21 March 1908). Some sources also state that on 29 March, he became the first to take a passenger into the air, Leon Delagrange. Later in 1908, on 30 October, Henri Farman went on to make the first cross-country flight in Europe, flying from Châlons to Reims. In the same year in July, he travelled to the United States to engage in exhibition flying.

As well as a love of speed-related activities, Henri and his brother Maurice (1877–1964) used the fame they achieved in racing to support their flying activities. Henri used his motorcar racing fame to become a seller of cars and then used the money from these sales to begin flying. Without any formal training as an engineer, Henri had an intuitive knack for mechanical things, encouraged by his experience with motorcars.

Henri's first venture into aviation as a business took place in 1909 when he opened a flying school at Châlons-sur-Marne at which George Bertram Cockburn was the first pupil. In the same year, he made further record-breaking flights of 180 kilometres in just over 3 hours (at Reims on 27 August) and 232 kilometres in 4 hours 17 minutes and 53 seconds (at Mourmelon-le-Grand on 3 November). At the end of 1909, Farman fell out with Gabriel Voisin because Voisin had sold an aircraft that had been built to Farman's specifications to J.T.C. Moore-Brabazon. Farman started manufacturing aircraft to his own design. The first of these, the Farman III, was an immediate success and was widely imitated.

In partnership with his two brothers Maurice and Richard (1872-1940), he built a highly successful and innovative aircraft manufacturing plant (the Farman Aviation Works; Avions Farman: Société Henri et Maurice Farman), and their 1914 model was used extensively for artillery observation and reconnaissance during World War I. After the end of the war, the Farman Aircraft Company's *Goliath* was the first long-distance passenger airliner, beginning regular Paris-London (Croydon Airport) flights on 8 February 1919.

Henri was made a chevalier of the French Légion d'honneur in 1919 and stayed in the industry until, along with Maurice, he retired in 1937, when the French Popular Front government nationalised the aircraft industry. Farman's company becoming part of the Société Nationale de Constructions Aeronautiques du Centre.

Henry's brother, Richard, like his brothers, was educated at home and was allowed an unusual amount of freedom to follow his personal interests. He was trained as an engineer, and he set up Paris' largest motorcar agency, the Palais de l'Automobile, dealing in Delauney-Bellevilles, Panhard-Levassors, and Renaults. Together with his brothers, he wrote *The Aviator's Companion* (published in 1910), describing their early flying achievements.

He became an electrical engineer and built the first electric trams in Brazil, in Rio de Janeiro. He was also an author of many technical works on engines, an aviator from 1914 to 1918 and was the director of the brothers' aircraft factory in Lyon. Richard Farman tended to concentrate on the business side of manufacture at the Société Anonyme des Usines Farman and their airline, which became part of Air France in 1933. In spite of his scientific knowledge and techniques, he dealt with administrative and commercial services. When he was aged 65, he retired from the industry after the French Government's nationalisation of the aircraft industry in 1937 ended the company's independence. Richard Farman died in Paris 31 January 1940.

The third brother, Maurice, was a champion cyclist and began racing Panhard motorcars, winning the Pau Grand Prix in 1901. In May 1902, he won the 'Circuit du Nord' race from Paris to Arras and back and competed in that year's race from Paris to Vienna. Maurice's interest later turned to powered flight, and in 1909, he set world endurance and speed records. In 1912, he merged his business with his Henri's aircraft factory and continued working with his brothers on the manufacture of aircraft. Maurice Farman died in Paris in 1964.

In 1941, the surviving Farman brothers (Henri and Maurice) reestablished their firm as the Société Anonyme des Usines Farman (SAUF), but only three years later it was absorbed by Sud Ouest. Maurice's son, Marcel Farman, reestablished the SAUF in 1952, but his efforts proved unsuccessful and the firm was dissolved in 1956. The Farman brothers designed and built more than 200 types of aircraft between 1908 and 1941, and also built motorcars until 1931.

Family wealth enabled the Farman brothers to pursue their early aviation interests with gusto. Buying a Voisin plane allowed them to quickly start setting records, demonstrating their skill and building their reputation as early movers. This was cemented with their publication of their early flying achievements which no doubt would have been readily received in the European aviation community. The growing reputation as flyers would have helped them in the establishment of their flying school. Wealth and early success allowed them to avoid the challenges of start-up costs faced by less wealthy aviation entrepreneurs, giving them an early competitive advantage. They experienced the downside of collaboration with Voisin when their aircraft specified design was sold to a competitor without their approval. However, having money meant they were able to quickly move into their own manufacturing, combining the strength of Richard's business acumen with Maurice and Henri's design capability.

5.6 Marcel Dassault

Marcel Dassault (born Marcel Bloch; 1892–1986) was a French aircraft industrialist. Born on 22 January 1892 in Paris, he was one of four brothers, his father working as a doctor. Both of his parents were Jewish, and he was educated at the Lycée Condorcet in Paris. After studies in electrical engineering, he graduated from the Breguet School and Supaéro, which at that time was a technical school specialising in vocational training in electrical engineering and left there when he was 19 years old (Dassault, 1971, p. 17). At the latter school, Bloch was classmates with a Russian student named Mikhail Gurevich, who would later be instrumental in the creating of the MiG aircraft series in the Soviet Union.

Dassault first worked as a trainee in a motorcar factory and then went to the École Superieure d'Aeronatique. He worked at the French Aeronautics Research Laboratory during World War I and invented a type of aircraft propeller which was subsequently used by the French Army during the war. It was through this work in wartime France that he gained a knowledge of production techniques and management and he later formed a small company with Henri Potez, the Société d' Etudes des Aeronautiques. During the downturn in the industry in the early 1920s, he concentrated on property development and then reformed his company in 1928 (Dassault, 1971, p. 37). This company was known as the Société des Avions Marcel Bloch, which produced its first aircraft in 1930.

In 1935, Dassault and Henry Potez entered into an agreement to buy Société Aérienne Bordelaise (SAB), but in 1936, the company was nationalised as the SNCASO. Unlike the Farman brothers, Dassault agreed to become the delegated administrator of the minister for air.

During Nazi Germany's occupation of France, the country's aviation industry was virtually disbanded, except for the compulsory manufacturing, assembly and servicing of German designs. In October 1940, Dassault refused to collaborate with the Germans occupiers at Bordeaux-Aéronautique and was imprisoned by the Vichy government. In 1944, the Nazis deported him to the Buchenwald concentration camp as punishment for refusing to cooperate with their regime. In the meantime, his wife was interned near Paris and he stayed in the concentration camp until it was liberated on 11 April 1945. By the time of his return to Paris he was crippled to such an extent that he could barely walk and was advised by his doctors to settle his affairs, as they did not expect him to recover his health.

It was at this time that he changed his name from Bloch to Bloch-Dassault and, in 1949, to simply Dassault. Dassault had been the alias used by his brother, General Darius Paul Bloch, when he served in the French resistance during the war and is derived from *char d'assaut*, French for 'battle tank'. In 1971, Dassault acquired Breguet, forming Avions Marcel Dassault-Breguet Aviation (AMD-BA).

In 1919, Bloch married Madeleine Minckes, the daughter of a wealthy Jewish family of furniture dealers. They had two sons, Claude and Serge. He resigned as managing director of Dassault when he was elected as a deputy to the National Assembly in 1951, but he remained on the board of directors (Dassault, 1971, p. 119). He died in Neuilly-sur-Seine in 1986. Serge Dassault, Marcel's son, became CEO of Avions Marcel Dassault, which was restructured as Groupe Industriel Marcel Dassault, reflecting its broader interests. In 1990, the aviation division was renamed Dassault Aviation.

Although Marcel Dassault came later than the initial pioneers of aviation in France, he had some characteristics in common with them. He had a strong engineering education and background, as well as a love of designing and inventing new gadgets. Although he initially worked for other companies, he eventually became restless with a conventional career in engineering and struck out on his own. During World War II, despite considerable setbacks, he remained committed to aviation and was determined to return to it once the war was over and he could return to France. In addition, he had considerable managerial skills, which made him stand apart from other early aviation pioneers.

Dassault's career deviated somewhat from that of other aviation entrepreneurs. His work with the French Aeronautics Research Laboratory in World War I connected him early on to the government and military, which ultimately contributed to sales of his aircraft and propellers and later to his position as administrator for the minister of air. His refusal to cooperate with the Germans under occupation signalled him out as a man of principle and his dedication to aviation meant he continued, against the odds, to pursue his passion after World War II, acquiring Breguet late in his life. He also dealt with the blows of the early 1920s in a different way to his rivals, choosing to diversify into real estate developments rather than the manufacture of other products. This proved to be a very successful commercial strategy and he was one of the more commercially successful of the aviation pioneers during the 1920s.

5.7 Conclusion

A number of the early French pioneers were important at creating aviation manufacturing before World War I. During the war, the companies these men formed created aircraft on a massive scale, enabling the French military to run the world's largest air force. A number of these companies arose as a result of partnerships with each other. A succession of shocks to the industry in the 1920s, 1930s and 1940s led to the disappearance of many of the pioneers from the industry.

The post-World War I collapse, the Great Depression, nationalisation in 1936 and the German occupation all put pressure on the French industry and led to the failure of a number of the original French companies. A number of French engineers, however, were to enter the industry at this time and were later to resurrect the industry in the post-World War II era.

The French pioneers tended to have strong engineering education and experience in their backgrounds, although they also tended to become frustrated with working for established companies and very early on branched out on their own. Flying instruction and prize money were used to finance the initial manufacturing concerns a number of them stayed in the industry for a considerable length of time, even into the post-World War II period. A number of them were keen on speed-related sports besides flying, such as cycling and the racing of motorcars. Funding of the initial companies tended to come from private sources, although the use of prize money and profits from flying schools and exhibition flying were common, as it had been in the United States.

Although the French industry is generally regarded as being unsuccessful in the 1920s and 1930s compared with developments taking place at the time in the United States and the United Kingdom, the industry did survive and in the post-World War II period became one of France's preeminent manufacturing sectors. Many of the pioneers of the French industry by this time had left the industry but it is notable that some, like Dassault and Saulnier, were amongst the most important managers of the post-war industry. It was not in France, however, that the most enduring legacy of the aviation pioneers was to occur. Instead, it was in the United Kingdom where a number of the industry's early flyers were able to stay in the industry for very long periods, and it is to this industry that we now turn in the next chapter.

References

Bakewell, P. (2010). On pioneering wings in France. *Southwest Review*, 95(3), pp. 479–500.

Barros, A. (2011). Razing babel and the problems of constructing peace: France, Britain and air power: 1916–1928. *English Historical Review*, 126(518), pp. 75–115.

Baughen, G. (2018). *The Rise and Fall of the French Air Force: French Air Operations and Strategy 1900–1940*. Stroud, UK: Fonthill Media.

Chapman, H. (1990). *State Capitalism and Working-Class Radicalism in the French Air-Craft Industry*. Berkeley, CA: University of California Press.

Chadeau, E. (1988). *De Blériot a Dassault: Historie de L'Industrie Aeronautique en France 1900–1950*. Paris: Fayard.

Dassault, M. (1971). *The Talisman: The Autobiography of Marcel Dassault Creator of the Mirage Jet*. New Rochelle, NY: Arlington House.

Elliot, B. A. (2000). *Blériot: Herald of an Age*. Stroud, UK: Tempus.

Gibbs-Smith, C. H. (1953). *A History of Flying*. London: Batsford.

Higham, R. (2012). *Two Roads to War: The French and British air arms from Versailles to Dunkirk*. Annapolis, MD: Naval Institute Press.
Lockwood Marsh, W. (1935). Tribute to a pioneer: The early trials and evolution of the Blériot monoplane. *Aircraft Engineering*, 7(77), pp. 159–160.
Sanger, R. (2008). *Blériot in Britain*. Tonbridge Kent, UK: Air Britain (Historians) Ltd.
Truelle, J. (1969). La Production Aeronautique Militarie Francaise Jusqu'en Juin 1940. *RHGGM*, 73, pp. 75–110.

6 The British pioneers

A.V. Roe, the Short brothers, Robert Blackburn and Frederick Handley Page

6.1 Introduction

In the pre-World War I period, flight also spread to the United Kingdom, where a number of British pioneers, inspired by the example of the American and French aviators, took to the air. A number of these pioneers formed companies before the war broke out, as was the case with the French companies. At the outbreak of World War I, these companies quickly turned into large-scale enterprises, mass producing substantial numbers of military aircraft. In the interwar years, the industry struggled, but unlike many of the French and American pioneer aviators who had left the industry, many British aviation pioneers stayed in the industry, surviving the interwar period and lasting into the 1950s.

Most of the British pioneers were quite young when they formed their companies, and generally had backgrounds in engineering and technical school education. Like the American and French pioneers, it was common for the British pioneers to have experience in the motor trades, bicycling and light engine manufacturing.

In this chapter, therefore, a sample of British aviation pioneers are described, along with their backgrounds and behaviours as entrepreneurs. In particular, A.V. Roe, the Short brothers, Robert Blackburn and Frederick Handley Page are looked at.

6.2 A.V. Roe

In England, one of the first important pioneers of heavier-than-air flight was Alliott Verdon-Roe (1877–1958). He was born in 1877 in Manchester, the fourth child of a family of three girls and four boys of Edwin Hodgson Roe, a doctor, and Annie Sophia Verdon. Roe was educated at St Paul's School, leaving at the age of 15 to spend a year working with an uncle in British Columbia in Canada before returning to England in 1893 to become an apprentice with the Lancashire and Yorkshire Railway Locomotive Works, located near Manchester. After completing his apprenticeship, he went on to study marine engineering at King's College, London and subsequently in 1899, he joined the British and South African Royal Mail Company, later being employed in the motorcar industry with Brotherton-Crokers Ltd. as a draughtsman.

Like many of the early aviator/entrepreneurs in England, Roe, therefore, had a technical diploma education as well as experience as an apprentice in the engineering trades. Roe was also the type of man who changed employment often, easily becoming bored with jobs that he found mundane. He did this right up until he finally settled down to building and flying aircraft, a pursuit he was to find endlessly fascinating. This suggests that Roe adopted an effectuation exploratory reasoning approach to his work that was more concerned with the design aspects and practical pursuit of flight rather than a desire to operate a business. This approach encouraged him to explore different jobs and industries around a common technical theme, which allowed him to develop skill sets useful for the pursuit of his 'flight' hobbies. This exploratory approach, focused on recognising and acting on opportunities, resulted in a transitory early career in engineering. Roe, therefore, had an ultimately unsteady early career but was able to experience considerable success in aviation and is attributed by some to have designed, constructed and flown the first British-designed and British-built aircraft to rise from the ground under its own power.

Roe is generally regarded as being one of the most inventive of the early builders of aircraft, with an inquisitive mind which was used early in his life to study and make aircraft models, bike riding (winning a number of competitions) and inventing various products (see Figure 6.1; Driver, 1997, p. 19; Roe, 1939, p. 22).

Figure 6.1 A youthful looking A.V. Roe

This meant that like a number of his French, American and British contemporaries he enjoyed hobbies that involved the pursuit of speed and remained throughout his life a compulsive inventor. In doing so, he first patented one of his inventions when he was only 13 years old. Later in life when he wrote about his work and the inventive spirit he had, he stated, "I started inventing young. In fact, I was only thirteen when I took out my first patent" (Roe, 1939, p. 162).

This inventiveness did have a more practical result for him. Roe in 1907 entered and won a model aircraft competition, defeating 200 competitors in the process, winning a prize of £75. He used this money to construct a full-sized aircraft, which was a copy of his winning model aircraft. In May 1908, he took the next step and fitted a more powerful engine to his machine and made a number of short flights. This is regarded as being the first occasion when a British-designed and built aircraft had risen from the ground under its own power. He claimed 8 June 1908 as the date of his first short flight. At the same time, he was embarking on these activities he attempted to remain in touch with developments elsewhere in the fledgling industry, and to that end, in 1910, he became a member of the Royal Aero Flying Club.

Roe's aim at this time was not only to fly aircraft but also to design and develop a continuing succession of improving models. As an entrepreneur focused on pursuing the opportunities flight promised, his interests lay in exploration, pushing the technology to the limits and improving on previous failures (Brousseau et al., 1996). To this end, he was determined to begin aircraft manufacturing, and in January 1910, he enlisted the help of his brother, H.V. Roe, who at the time ran a manufacturing firm in Manchester. The two brothers together founded A.V. Roe & Co. on 1 January 1910 and split the work of the company between them. In doing so, A.V. Roe focused on inventing, experimenting and constructing new models of aircraft, while his brother H.V. focused on the finance and management of the firm (Driver, 1997, p. 64). Initially, the firm made few private sales, but from 1912 onwards, they began to sell aircraft, first to the War Office and then later to the Admiralty. Important to the success of the firm was the designing and construction of the Avro 504 in 1911, which in an improved form became one of the best-known military aircraft of World War I, serving for many years as the main basic pilot trainer of the British military (Holmes, 2004).

To finance the establishment of this small company the brothers raised money from their immediate family, who insisted that one of the conditions of this assistance to them was that H.V. Roe be made the managing director of the company (Ludovici, 1956). This was because the family members regarded him as having greater business acumen, and there was a suspicion that A.V. Roe would lose the money if he was left in charge. In this example of a company start-up, the difficulty of combining management acumen with an innovative spirit and skill was solved by having each of the two brothers specialise in their own areas of expertise. A.V. pursued the technical opportunities wherever they lay, which were well within his control and interest, while his brother followed a steady structured expert career path that favoured a causation approach (Politis, 2005, Sarasvathy,

2001). This solution was, however, not a permanent one, and, finally, the brothers fell out over how the firm should be run. H.V. Roe left the company to join the military in August 1917, serving as an aircraft gunner on the Western Front. This crisis forced A.V. Roe to take on a more active management role.

In working for his company A.V. Roe made considerable technological progress and designed and built first the Avro 500 and later the Avro 504. These designs led to considerable military sales. Sales of the Avro 500 aircraft before World War I meant that brothers were able to find additional investors, and in January 1913, the company was made a limited company (Driver, 1997, p. 78). James Grimble Grove was the main new investor in the company, and he received 10 per cent preference shares, with the Roes holding the rest as ordinary shares. Later in raising finance A.V. and H.V. Roe issued 10 per cent preference shares to investors, while they themselves "took payment in ordinary shares, which were to receive two thirds of the remaining profit" (Roe, 1939, p. 140).

Orders from the War Office for the Avro 504 began in April 1914 and during the war, the company expanded greatly. After the war ended the company struggled to deal with the slump in demand (Driver, 1997, p. 81). A.V. Roe attempted to diversify production in the face of the falling sales of aircraft, but he met with little success. In the face of the evaporating sales, Crossley Motors claimed a majority shareholding in the firm in May 1920 by buying out the shares of Roe's family and others, leaving A.V. Roe to stay on as minority shareholder (selling his own shares in 1928).

The firm's attempts in the early 1920s to diversify met with little success, and the company struggled, as most did at this time, until in 1928 a controlling interest in A.V. Roe & Co. was purchased by the Armstrong Siddeley Motor Company. A.V. Roe took the money from the sale of his shares in 1928 and, along with two other investors, bought an interest in the firm Saunders Ltd., a boat builder based in Cowes. The firm's name was later changed to Saunders-Roe and it combined the construction and sale of flying boats and marine craft. Again, A.V. Roe drew upon the controllable aspects of his life – his technical interests and expertise – to move across to a related industry to further experiment, develop and extend his previous knowledge and skills in the next stage of his career. Much of his early success, however, depended on the steady business skills of his brother. A.V. Roe was to remain president of Saunders-Roe for the rest of his life until he passed away in 1958.

A.V. Roe was a visionary, full of inventive ideas, who believed that mechanical flight could deliver tremendous world benefits. Roe also believed that speeds would increase with height over the earth and was always looking forward to what changes would occur in the future of aviation. His focus was very much on the benefits of the future and enjoying the unpredictability of the journey to get there (Sarasvathy, 2001). He possessed many of the skills that the aviator/entrepreneurs possessed, including an engineering education/experience, a passion for working on mechanical problems, a restless spirit when it came to pursuing stable employment in engineering, a love of speed associated sports, initial finance and support from his family, the winning of prize money, and a willingness to risk whatever capital he could lay his hands on. He was also an early member of the British Aero Club, allowing him to stay in touch with others in the industry. His one main crucial weakness was in management and financial skills, which he solved early on by utilising his very accomplished brother. Roe managed to hang on in the industry in the early 1920s when many others dropped out, but his presence in the industry was never as important from the 1920s onwards to the same degree as it had been in World War I. Unfortunately for the brothers' partnership, his desire to achieve his vision often conflicted with his brother's steadier approach to business. In later years when A.V. Roe sold his shares in A.V. Roe & Co., he was able to purchase and built up Saunders-Roe as more a niche, experimental company, rather than a mass constructor of aircraft, a type of firm probably much more suited to his inventive spirit (Holmes, 2004; Ludovici, 1956).

6.3 The Short brothers

At the same time that A.V. and H.V. Roe were forming their company, another team of brothers formed one of the United Kingdom's longest lasting aviation companies. Of these, Oswald Short (1883–1969) was born in Stanton by Dale, Derbyshire, on 16 January 1883. He was the fourth son and fifth child of Samuel Short, an engineer, and his second wife, Emma Robinson. Oswald Short's formal education ended on his father's death, but he trained as an engineer through self-help and hands-on instruction from his brothers Horace Leonard (1872–1917) and (Albert) Eustace (1875–1932).

Oswald Short first worked in the 1890s as an office boy in Derby and founded a balloon business in 1898 with his brother Eustace. In April 1901, the two brothers built their first passenger-carrying balloon at Hove and moved their factory to Maple Mews, Tottenham Court Road, London, around 1902. To begin with, the partnership's first sales were observation balloons for the government of India (October 1903), and from 1905 to 1906, the brothers established contacts in the Aero Club, moving their factory in June 1906 to Battersea. Further balloons were made and sold to early important aviation figures such as C.S. Rolls, J.T.C. Moore-Brabazon and T.O.M. Sopwith, and the brothers established a reputation for quality, becoming the club's official aeronauts (1907) and aeronautical engineers (1908). Rolls and Moore-Brabazon had been founders of the Royal Aero Club.

From 1907, Oswald and Eustace Short tried unsuccessfully to build heavier-than-air aircraft designed by club members, and in December 1908, Horace left his employment with Charles A. Parsons to form with the other two brothers a tripartite partnership, which was known as the Short Brothers, Aeronautical Engineers. Their rising reputation meant that in March 1909, Wilbur Wright chose the Shorts to build six Model A Flyers (the United Kingdom's first batch contract) and these were manufactured in the country's first purpose-built aircraft factory, Shellbeach, Isle of Sheppey. They were used to great effect, and in October 1909, Moore-Brabazon won the *Daily Mail*'s £1,000 prize for the first circular mile by an all-British aircraft in a Short No. 2 powered by an engine designed by Gustavus Green. The brothers' success was due to a hard-headed practical approach to cultivating business in the civilian market, which sometimes offended others. In 1911, the Shorts provided the Royal Navy with its first two aircraft and were to become the principal supplier of the Royal Navy Air Service in World War I, after which they were to have a long association with naval aviation and flying boats.

Oswald Short established a Shorts' seaplane factory at Rochester (1913–14) and the airship works at Cardington, Bedfordshire (1916). When his brother Horace passed away in 1917, Oswald became wholly responsible for aeroplane design. Oswald Short experimented with light alloy construction from about 1916 and pioneered and patented the smooth-skinned monocoque light alloy construction (stressed skin) in the United Kingdom with Silver Streak (1920). Later he also applied a non-corrosive form of the technology to flying boats and persevered with stressed-skin construction in the face of official indifference. He was able to keep the company going through the 1920s by building motorbus bodies until the Royal Air Force and Imperial Airways were willing to accept stressed-skin-made aircraft. The firm of Shorts was also able to make money at this time from licencing agreements to build aircraft in the United States, France and Japan.

The Short partnership was incorporated as Short Brothers (Rochester and Bedford) Limited in 1919 with Oswald Short as chairman and joint managing director, and he became sole managing director on Eustace's death and retained both posts when Shorts became a public company in 1935. The peak of Oswald's career as an aircraft builder was marked by contracts for Short Empire flying boats for Imperial Airways; Sunderlands and Stirlings for the Royal Air Force in the mid-1930s; and the incorporation of a subsidiary, Short and Harland, in Northern Ireland.

In 1935, Oswald Short married Violet Louise Blackburn, *née* Lister. His health failed under the stresses of running his company in the 1930s and in the early years of World War II. In the middle of the war, unhappy about the production levels of the company, the British Government expropriated the shareholders in 1943, and Oswald Short resigned as chairman and director of Shorts, accepting the honorary title of life president.

From 1943, Oswald Short lived quietly in retirement, disappointed about the government's takeover of his company. He lived on long after World War II, largely forgotten except to those in the industry itself, finally passing away in 1969. The firm was renamed Short Brothers Limited in 1977 in honour of the brothers, but the memorial to Short's own genius is the worldwide acceptance of stressed-skin construction. After being privatised in the 1980s, the company was taken over by Bombadier of Canada and the factories in Belfast still produce aircraft components for that company today.

Early on in his career Oswald Short was captivated by and moved into aviation, first through balloons and then through planes manufactured by the Short brothers' businesses. The Short brothers' work life was largely within businesses they created. Oswald's early contact with the Aero Club connected him to a number of emerging important aviation figures who recognised and promoted the quality of the Short brothers' work. This reputation was endorsed by the Wright brothers' contract, the winning of the *Daily Mail* prize and their appointment as principal suppliers to the Royal Navy Air Service. Finding new applications in related industries for his quality technology innovations and related licencing agreements demonstrates significant entrepreneurial ability, which kept the company afloat during the interwar years. Connections and quality work and the enduringly unique innovation of stressed skin made the Short brothers early leaders in establishing foundational knowledge that underpinned the development of the aviation industry.

6.4 Robert Blackburn

Another early pioneer in the British aviation industry was Robert Blackburn (1885–1955). Blackburn was born on 26 March 1885 at Kirkstall near Leeds, the eldest of three sons of George William Blackburn, an engineer, and his wife, Kate Naylor. Blackburn's career was influenced by his father, who was the works manager of a lawnmower and steamroller manufacturing firm (William Green & Sons), and by the world's first heavier-than-air flight by the Wright brothers in 1903. He attended Leeds Modern School and gained an honours degree in engineering at Leeds University, later becoming an associate member of the Institution of Civil Engineers in 1906 and getting practical experience at his father's firm. Unlike many of the other early British pioneers of aviation, Blackburn had an engineering degree, rather than a technical college background. His family background was, however, in engineering, like so many of the other British pioneers.

After graduation, Blackburn went to work for his father's company, but he considered the company's methods old-fashioned. In order to broaden his own knowledge of engineering, he went to the continent to study new techniques. In France, he saw Wilbur Wright fly, and this proved to be the catalyst for his future career. He began to design an aircraft in Paris during that year (1908), and when he returned home, he left his father's company. His father set him up with a workshop in Benson Street, Leeds, so that he could work on building aircraft, and it was here that the first Blackburn aircraft, a monoplane powered by a 35 h.p. Green engine, was built. He had it transported to Marske by Sea on the Yorkshire coast, where he taxied it along the sands at increasing speeds, eventually becoming airborne. Although the machine side-slipped into the ground and was wrecked, Blackburn had achieved his initial ambition and his second monoplane was more successful (see Figure 6.2). Within three years he had proved himself as designer, builder and pilot, and by June 1914, he had become a manufacturer, founding his own company, the Blackburn Aeroplane and Motor Company, as a limited company. The company had capital of £20,000 in £1 shares, with himself and a number of local flying enthusiasts as directors and investors. Prior to this he had made a series of individual planes,

Figure 6.2 Monoplane of Robert Blackburn at Filey

as well as operated a flying school, the latter being the main source of income for him (Jackson, 1968, p. 6).

He was twice married, first to Jessica Tryphe Thompson on 31 October 1914; they had three sons and two daughters, although this marriage was dissolved on 23 March 1936. His second wife was Phyllis Margaret Kirton, whom he married on 9 April 1936; they had two daughters. Both wives survived him.

Blackburn's genius for organising large-scale production showed itself during World War I when the company used subcontracting arrangements to undertake a wide variety of aircraft work to fulfil Admiralty and War Office orders. The main aircraft that he built for the government were the B.E.2s, which were the mainstay of British aviation in the early years of the war, but at the same time, Blackburn worked on the development of a range of floatplanes.

During 1915, Blackburn undertook two important developments for his company – the acquisition of an airfield and slipway at Brough, by the River Humber, and a decision to specialise in naval aircraft design and manufacture. In the 1920s, his company struggled to survive and did so by producing a range of products such as nuts, bolts, turnbuckles and bodies for cars and buses. Instead of permanently changing to the production of these products, Blackburn tried to stay in the aviation industry, and this he did so by developing a single-engine torpedo bomber in the early 1920s, which he sold successfully to the British Government as well as overseas. In particular, at the Greek government's invitation in 1925, a national aircraft factory near Athens was opened by his company. For this, he was awarded the Gold Cross of the Order of the Redeemer by the Greek Government.

Through the 1920s and 1930s Blackburn continued to specialise in the production of naval aircraft and during World War II, Blackburn's Brough works, to which all manufacturing had been transferred in 1936, built 1,700 Fairey Swordfishes and 700 Barracudas in addition to its own Skuas and Rocs for the Royal Navy, plus Bothas for the RAF, and co-operated with the Clydeside shipbuilders Dennys of Dumbarton in building Sunderland flying boats.

After World War II, Blackburn repeated his post-World I strategy of producing non-aviation products, such as light alloy roof trusses, agricultural implements and domestic appliances. A 1948 amalgamation with General Aircraft led to the production of the Beverley heavy freighter for the RAF, with Blackburn staying on at the company until his death in 1955. The company did not long survive the death of its founder, and in 1960, it was bought out by Hawker Siddeley.

Robert Blackburn, as the founder and managing director of his company, sought to retain the independence of his company, even though it lacked the success that a number of the British companies did in the design and manufacture of aircraft. Blackburn himself seems to have exhibited a considerable ability to withstand the downturns in demand that occurred after each of the world wars through the short-term production of a variety of manufactured products and maintained a long determination to stay in the industry and reject the temptation to branch off into other fields. British aviation had a number of other pioneers who exhibited similar characteristics, and a number of British firms maintained their independence due to the determination of their founders.

Blackburn's commitment to the usage of his own designs and his manufacturing and flight skills shaped his career. Like other entrepreneurs, his flight school connected him to local flying enthusiasts, some of whom financed his ventures as well as provide him with an additional stream of income in the early years. He was gifted in starting and running a business, even through tough times, and his strong ability to adapt and manage multiple stakeholders and concerns separated him from other entrepreneurs, who often struggled with the operational side of running a business. Specialisation in naval aviation allowed him to carve out a niche and build up resources and expertise that sustained him through the tough times. Throughout the tough times when he was manufacturing non-aviation parts, he continued to undertake aviation innovation; he was able to identify where future aviation demands lay and to quickly capitalise when the environment became receptive to his inventions.

6.5 Frederick Handley Page

Another of the early pioneer aviators was Frederick Handley Page (1885–1962), who was born in Cheltenham in 1885, the second son of Frederick Joseph Page, a furniture maker and member of the Plymouth Brethren. Handley Page was educated at Cheltenham Grammar School, and in 1902, against his parents' wishes, he moved to London to study electrical engineering at the Finsbury Technical College. In 1906, he qualified and was appointed to be the head designer at Johnson &

Phillips Ltd., an electrical engineering company in London (Barnes, 1976; Handley Page Ltd., 2012; Pritchard, 1962).

The catalyst for Handley Page setting up his own business was his dismissal in 1908 by Johnson & Phillips Ltd., which was unhappy with his undertaking aviation experiments on their premises with the company's equipment. This event could be construed as a crisis for Page, who responded creatively by setting up his own aircraft construction business, despite being only 24 years old and having little money. Handley Page Ltd. was founded in 1909 (the first British company registered specifically to manufacture aircraft) and had an authorised capital of £10,000 through the creation of 500 £20 shares, although initial capital was, however, far smaller than that. Handley Page became the managing director of the company and the board was made up of himself, his brother and a couple of local engineers (Driver, 1997, p. 90). He earnt extra money at this time by doing evening teaching at Finsbury Technical College and then, in 1911, he joined the Northampton Polytechnic Institute (later Northampton Engineering College and then the City University) as a lecturer in aeronautics. Like A.V. Roe and Robert Blackburn, Handley Page enjoyed his independence and his experimentation with aviation was not confined by being employed in an adjacent industry in a senior position. It was inevitable that this focus on his passion was not going to lead to a linear career path and the need to finance his start-up encouraged him to take up two jobs in aeronautics but in two different areas (as teacher and as business owner).

He built his first passenger-carrying monoplane and flew it across London in July 1911, and in the following year, another two-seater was able to compete in military trials on Salisbury Plain (Barnes, 1976). His firm was to receive its first contract to supply aircraft from the War Office in 1913 and later during the war, the Admiralty ordered twin-engine bombers from Handley Page (Driver, 1997, p. 64). This was followed in 1918 by the design, construction and sale of even larger four-engine machines, which were to be used to bomb Berlin (Barnes, 1976). Strong growth in demand in the latter years of the war meant that by December 1917, the company employed 1,120 workers and by early 1919 2,500, and delivered 211 aircraft in 1918 and a further 103 in 1919 (Driver, 1997, p. 64).

Success late in World War I prompted Handley Page to do what a number of aviation manufacturers were later to do: to offer shares in his company to the public, after the successful creation of aircraft types for which there was a demand. In 1919, members of the public took up the shares, despite the speculative nature of the industry in peacetime, paying 21s for each of 500,000 7 per cent cumulative preference shares of £1. The ordinary shares (150,000 £1 shares) were allotted to the original shareholders, mainly to Frederick Handley Page himself. Unfortunately, the timing of this public share issue was at a risky time for the industry, as demand for aircraft slumped in 1920, leaving the company in an exposed position.

In the face of this decline in demand, Handley Page attempted to diversify the interests of his company, and in 1919, Handley Page Transport Ltd. was set up to promote a civil air service to the continent (it was later sold to Imperial Airways in 1924 after consistently making a loss) (Handley Page Ltd., 2012). Handley

Page also attempted car assembly and the sale of aircraft overseas; however, both of these two ventures also lost money. Aircraft manufacturing incurred massive losses, and Handley Page lost control of his company, it being governed by a chairman appointed by the company's creditors. Handley Page was retained by the board of the company as managing director as it was thought he still possessed substantial aeronautical construction, design and management skills, as well as owning a number of the patents that helped the company to survive.

The company itself managed to eventually trade itself out of insolvency through getting income from sales of aircraft from the Aircraft Disposal Company, which it had investments in, and from the sale of the transport firm to Imperial Airways in 1924. In addition, royalties from the slotted wing patented by Frederick Handley Page also helped the firm. By 1926, the company had discharged its indebtedness to the major creditors and its preference shares were then revalued downwards (Driver, 1997, p. 70). Gradually through the 1920s, the sale of heavy bombers and civil aircraft recovered, and the company was profitable by the late 1920s. Profits were later to rise from the mid-1930s onwards when rearmament began, and the company did not raise capital at this time from the sale of shares, as a number of other companies did, perhaps because of the difficulties experienced in the early 1920s.

Handley Page's company focused on building heavy bombers in both world wars and large passenger aircraft in the interwar years (Handley Page Ltd., 2012). The firm built airliners for Imperial Airways at this time, but orders were too few to guarantee large profits, and no dividends were paid to ordinary shareholders throughout the 1920s. Eventually, during the late 1930s and during World War II, the company returned to profitability, producing the Hampden and later Halifax bombers. The company produced the first post-war civil airliner, the Hermes, and later the Victor bomber, but by the 1950s, the viability of the small independent aircraft producer was under serious threat (Handley Page Ltd., 2012). Handley Page obstinately refused to give in to government pressure to merge with other firms in the industry, and at the time of his death in 1962, his company was one of the few remaining independent aircraft manufactures left.

The takeover of the organisation in the early 1920s could be defined as a second crisis event for Handley Page, which resulted in higher order strategic thinking on his part. It could be argued that during this period Handley Page shifted from a focus on the long-term vision to a focus on what was needed to survive suggesting a shift from a spiral career path (common in entrepreneurs) to a linear/expert career path (common to non-entrepreneurs). Whether this was by personal choice or due to pressure from the appointment of an outside chairman by the creditors to the company is an interesting consideration. When the bottom dropped out of the aviation market in the early 1920s, he was therefore not well positioned to re-adopt an exploratory approach to the business, which ultimately led to him being kept on at the firm for his technical expertise and his control of important patents. When Handley Page Limited's fortunes changed, he was, however, well positioned to cope with the liabilities of newness. His substantial aeronautical aviation skills and patents contributed towards him re-establishing his own and the company's prestige enabling him to maintain the independence of the company until his death.

Handley Page, in addition, was a pioneer of technical education and was to remain involved in it throughout his adult life, both at the technical college and university levels. He used his teaching early in his career to support himself while he worked on his various aircraft projects, but he also remained directly involved in the development of technical education even after he became the managing director of his company. In addition, he also favoured research in technological developments and often later in life wrote and spoke about this aspect of the industry. Very notable about his career was his steadfast refusal to merge his company with others in the late 1950s, an illustration of his very strong determination to run things his own way, and a common characteristic of many of the aviation pioneers/entrepreneurs (Fearon, 1978).

6.6 Conclusion

Like their American and French counterparts, the British pioneers often sought to enter the manufacturing of aircraft, although in the wake of the collapse in demand post-World War I also struggled to survive. Despite various bankruptcies and takeovers, A.V. Roe, the Short brothers, Frederick Handley Page and Robert Blackburn all managed to remain in the industry, a product of their ability to take advantage of the limited military sales of the RAF in the 1920s and early 1930s. Handley Page was the only one to develop a significant presence in the developing commercial markets of the 1920s and 1930s, and even this was to a degree a by-product of the company's wartime development of heavy bombers. The others tended to survive by producing non-aviation products in the early 1920s and then by specialising in the production of types of military aircraft; Blackburn in naval aircraft and A.V. Roe and the Short brothers in floatplanes and flying boats.

Despite the differences with their American and French counterparts, the British pioneers were similar in that they tended to be young, adventurous, naturally inventive types, often with technical college educations (Robert Blackburn being a rare exception with an engineering degree), who had trouble settling into conventional engineering trades careers. Each also exhibited a dogged attachment to the industry riding out as best they could severe downturns in the early 1920s and early 1930s. Roe, Blackburn and Handley Page all managed to survive in the industry into the 1950s, an indication of their strong commitment to aviation and the determination to which they pursued their interests.

References

Brousseau, K. R., Driver, M. J., Eneroth, K. and Larson, R. (1996). Career pandemonium: Realigning organisations and individuals. *The Academy of Management Executive*, 10(4), pp. 52–66.

Driver, H. (1997). *The Birth of Military Aviation: Britain, 1903–1914*, Vol. 4. Woodbridge, UK: Boydell & Brewer Ltd.

Fearon, P. (1978). The vicissitudes of a British aircraft company: Handley Page Ltd. Between the wars. *Business History*, 20(1), pp. 63–86.

Handley Page Ltd. 2012 (1950). *Handley Page Forty Years on 1909–1949*. Stroud, UK: Fonthill Media.

Holmes, H. (2004). *Avro: The History of an Aircraft Company*. Ramsbury, UK: Crowood Press.

Jackson, A. J. (1968). *Blackburn Aircraft Since 1909*. London: Putnam.

Ludovici, L. J. (1956). *The Challenging Sky: The Life of Sir Alliott Verdon-Roe*. London: Jenkins.

Politis, D. (2005). The process of entrepreneurial learning: A conceptual framework. *Entrepreneurship Theory and Practice*, 29(4), pp. 399–424.

Pritchard, J. L. (1962). H. P. Sir Frederick Handley Page, C.B.E. *The Aeronautical Journal*, 66(624), pp. 737–742.

Roe, A. V. (1939). *The World of Wings and Things*. London: Hurst & Blackett.

Sarasvathy, S. D. (2001). *What Makes Entrepreneurs Entrepreneurial?* Charlottesville, VA: Darden Business Publishing.

7 The captains of industry

George Holt Thomas and Thomas Sopwith

7.1 Introduction

In its earliest years, most of the aviation manufacturing enterprises were very small-scale organisations, mainly made up of firms that consisted of design offices and workshops (Driver, 1997; Fearon, 1969). Even in the interwar period, aviation manufacturers tended to be of relatively small scale. World War I, however, transformed the industry into one of very massive scale (Driver, 1997; Fearon, 1969). The aviation pioneers struggled to increase production at this time, overcoming a number of difficulties, including the lack of widespread knowledge of working with the new technologies. During World War II aviation manufactures, again, struggled with the need to transform their relatively small enterprises into large-scale producers.

In each country, start-up entrepreneurs struggled with this change in the circumstances of the industry. Once World War I ended, the industry reverted back to a much smaller size and those substantial organisations not able to transition to smaller sizes often left the industry.

The purpose of this chapter is to see how two entrepreneurs (George Holt Thomas and Thomas Sopwith) coped with these changes of rapid expansion and then contraction. In the former case, George Holt Thomas was one of the few older initiators of the industry and decided to exit the industry in the 1920s. The latter, however, stayed in the industry pretty much longer than any of the pioneers and as late as the 1970s was still directly involved in the industry (Driver, 1997; Edgerton, 1991; Fearon, 1969).

7.2 George Holt Thomas

George Holt Thomas (1870–1929) was the most prominent industrial leader in the British aviation industry in its early years. He came from a newspaper background (his father owned a newspaper business) and knew the Farman brothers, whose parents were also in the newspaper business. Holt Thomas formed the Aircraft Manufacturing Company (often known as Airco) in 1912, and by November 1918 at the end of World War I, Holt Thomas' company was the largest manufacturer of aircraft in the United Kingdom. Despite this success, in March 1920 Holt Thomas sold Airco to the Birmingham Small Arms Company (BSA), which closed down its aviation activities and sold off what it could. Holt Thomas provides a good case of an early successful manufacturer of aircraft in the United Kingdom who struggled during the slump in demand for aircraft after World War I and who subsequently went on to fund alternative ventures that went on to success. Holt's career could be largely described as linear and within his control.

Holt was born on 31 March 1870 at Hampton House, Stockwell Private Road, in south London, the seventh son of the founder of the illustrated weekly *Graphic* and *Daily Graphic* newspapers, and his wife, the daughter of marine artist John Wilson Carmichael. Holt Thomas was educated privately at home and later attended King's College School, London, before spending two years (1888–90) at Queen's College, Oxford. He ended up leaving Oxford without completing his degree and instead joined his father's newspaper business. In 1894, he married Gertrude Hesley; however, the marriage was childless. By 1906, Holt Thomas had made a name and fortune as director, then general manager, of his father's newspapers and by founding his own illustrated weekly, *The Bystander*. In these journals, plus his *Empire Illustrated* magazine (1908–14), he vigorously advocated for the production of British goods for the British people at home and abroad. He was also a founder of the Association of British Motor Manufacturers, which sought to have the British Government impose an import duty on sales of foreign cars in the United Kingdom.

During 1906, however, Holt Thomas abandoned his active interest in the newspaper business and by 1920 had made a much greater name – and fortune – in the aviation industry. He was quick to recognise that the efforts being made by the aviation pioneers would soon have both commercial and military consequences, and felt that the United Kingdom should have a place in aviation. As early as April 1909, he claimed that the sum already spent on military aviation by Germany was about £400,000, by France about £47,000, and by the United Kingdom a mere £5,000. He was equally quick to recognise in the French pioneers Henri Farman and his brother Maurice allies well able to benefit his own and British interests. In April 1910, as a result of his friendship with the Farmans, Holt Thomas engaged a skilful pilot, Louis Paulhan, to compete for – and win – the prize of £10,000 put up by his friend Lord Northcliffe, the owner of the *Daily Mail*, for a successful flight from London to Manchester. Holt Thomas, himself, was important in inspiring Northcliffe's interest in aviation, and the prize was so great because no one had hitherto flown anything like that distance.

Holt Thomas was unlike many of the other founders of aviation manufactures in that he was an established businessman who had no real commercial need to enter the industry. Instead, he became interested in the industry and was older than most of the other founders at this time. He did have one advantage over his younger contemporaries in that he was able to draw on larger sources of financial capital to establish his ventures.

In 1912, Holt Thomas formed Airco and planned to build Farman aircraft at Hendon under licence. These aircraft were used as trainers in large numbers by the Royal Flying Corps and became known as Longhorns and Shorthorns (Edgerton,

1991). He remained as fervently patriotic as in his newspaper days, urging the government to take a more active interest in developing the United Kingdom's infant aviation industry. Holt Thomas was a practical man and despite his patriotic attachment to things British, he recognised the superiority of French airframes and engines at the time. He, therefore, obtained licences to manufacture the French Gnome and Le Rhône engines, which were widely used in British aircraft during World War I.

One of the keys to his success in business was his employment of high-quality personnel, and in May 1914, he invited Geoffrey de Havilland to join Airco, one of the most gifted aircraft designers in the United Kingdom's aviation history. Before this move de Havilland had been unhappily employed as an inspector and occasional test pilot at the Royal Aircraft Factory in Farnborough and jumped at the opportunity to resume his career as an aircraft designer. In the following six years, their partnership flourished, and de Havilland later recalled how glad he was to work for "a brilliantly clever as well as a kind and likeable man, a man who was far-sighted, extremely able and possessed a knowledge of business only equalled by his ignorance of engineering" (de Havilland, 1961, pp. 93–95). Holt Thomas and de Havilland formed an exceptional partnership, and during World War I, about 30 per cent of all fighters, bombers and trainers used by the United Kingdom and the United States were de Havilland designs.

At the time World War I ended in November 1918, Holt Thomas' Airco was the largest manufacturer of aircraft in the United Kingdom, operating a range of factories in different places, producing not only a series of conventional aircraft but also flying boats, airships, propellers and engines (Edgerton, 1991). His factories were generally equipped with the latest metal-working machinery, and at Hendon, he had installed a wind tunnel and a laboratory for material testing.

After the armistice, Holt Thomas hoped to keep his massive organisation in being and enjoy the same success in civil aviation as he had in military aviation. As early as October 1916, he had formed a company, Aircraft Transport and Travel, to manage this transition to commercial aviation. In addition, in March 1916, he founded the Society of British Aircraft Constructors to speak with one voice to government agencies to promote the industry's interests.

In May 1917, Holt Thomas told a meeting of the Aeronautical Society that military aviation should have been taken more seriously before 1914 and that the same mistake must not be made over civil aviation after victory was won. He foresaw internal airlines on a trunk route from London to Glasgow via Manchester, with feed-offs to Dublin and Belfast and links to the capitals of every continental country, including Russia, with eventual extensions to the United States and to New Zealand via India and Australia. Mail services and business and tourist flying, he argued, would all prove profitable.

To achieve these ends early in 1919, Holt Thomas headhunted men of the highest reputation in aviation – Sefton Brancker, Francis Festing and Mervyn O'Gorman – and on 25 August he started the world's first scheduled international air service. It operated daily, whether there were passengers or not, between London (Hounslow) and Paris (Le Bourget), using two of de Havilland's wartime designs until April 1920, when the DH 18, a biplane with seating for eight passengers, was brought into service: it was de Havilland's first aircraft designed purely as a civil transport. Holt Thomas was, however, a bit ahead of his time, and the service only lasted until November 1920, when it was shut down because of mounting losses.

By then Holt Thomas' hopes for a post-war boom in civil aviation had not come to pass as the technology in the industry at this stage could not support it. No aircraft at this time had the capacity to carry sufficient passengers, freight and mail over sufficient distances to be profitable, and government support for the research and development needed to produce such aircraft was not forthcoming. Faced with the bankruptcy of this company in March 1920, Holt Thomas sold Airco to BSA, which closed down its aviation activities and used most of its plant and equipment for other purposes.

Unlike a number of the other pioneer aviation entrepreneurs, Holt Thomas was of an older generation who by the time the post-war slump came along was ready to reduce his involvement in the industry and to sell out. By this time, he was in his 50s and was content to leave the industry in the hands of his younger colleagues. He still continued as a strong supporter of the industry but was not prepared to be one of those who struggled along in the 1930s like the much younger Robert Blackburn, A.V. Roe or Handley Page. In the United Kingdom, very few established businessmen were prepared to commit investment or time to the industry prior to World War I, and Holt Thomas was an important exception.

Holt Thomas did, however, continue to support the industry through the 1920s, although less actively. BSA sold some of the aviation assets it had acquired from Thomas Holt to Geoffrey de Havilland, who had decided to establish his own company and did so in September 1920, with Holt Thomas investing £10,000. Holt Thomas' investment made possible the creation of the company that would become world famous.

In addition, Holt Thomas published *Aerial Transport* (1920), along with a number of articles and lectures that supported the development of aviation. He retired to his country home at Hughenden, Buckinghamshire, where he took up the breeding of dairy cattle with something of the same enthusiasm he had formerly shown for newspapers and flight. His health became poor, and he died at the Queen Victoria Memorial Hospital, Cimiez, near Nice, France, on 1 January 1929, following an operation for throat cancer.

Holt Thomas was an important and driven visionary and advocate for the industry. His media background meant he was well positioned to lobby and promote to both the government and the public the promises of aviation. He was a man focused on action and starting things whether they be lobby groups or organisations and understood that to establish an industry required delivering quality reliable products, so he actively recruited and used the best in the business – be they employers, product or connections – to demonstrate the potential of aviation. Failing health and energy may have been a key reason why he personally withdrew from the industry, while continuing to support promising up comers like de Havilland.

7.3 Thomas Octave Sopwith

Holt Thomas' aviation empire was wound up in 1920 and was later to continue in some form as the de Havilland Company, but another organisation that went through a similar transition at this time was that of Thomas Octave Sopwith (1888–1989). Sopwith ran one of the other main manufacturers of aircraft in World War I in the United Kingdom but was much younger than Holt Thomas.

Sopwith was one of the most successful aviation entrepreneurs from any country and was one of the United Kingdom's original flyers. Sopwith was born in 1888 in Kensington, London, the eighth child and only son of Thomas Sopwith, who worked as the managing director of the Spanish Lead Mines Company of Linares in southern Spain, and his wife, Lydia Gertrude. Sopwith was educated at the Cottesmore School, Hove, Sussex, and, from 1902, the Seafield Park Engineering College at Lee-on-Solent. He left Seafield Park in 1905 without completing a diploma there but he did have a good practical grasp of basic engineering. Like so many of the other early pioneers/entrepreneurs, Sopwith pursued a range of adventurous hobbies including ballooning, motor racing at Brooklands, and sailing. Crisis came early to Sopwith in life when at the age of 10 years he accidentally killed his father with a shotgun while holidaying with the family in Scotland. Despite this terrible experience that haunted him for the rest of his life, Sopwith went on to found several companies that were instrumental in the development of the fledgling airline industry. He began his working life as a seller of motorcars, an extension of his early interest in riding motorcycles and racing cars, but soon turned his attention to flight.

In 1910, he joined the British Aero Club, purchased his first aircraft and after teaching himself to fly it, made his first flight. In the following years, he was able to win prizes both in the United Kingdom and in the United States, and by the end of 1910, he had set a British distance and duration record. In December 1910, with a flight of 169 miles in three and three-fourths hours, he won the £4,000 Baron de Forest prize for the longest flight of the year from the United Kingdom into Europe. He also won further prize money in the United States and used this money to set up the Sopwith School of Flying at the age of 24 in February 1912, and in June 1912, the Sopwith Aviation Company.

Aircraft sales to the Admiralty by Sopwith's company began in 1913, and the firm was reconstituted as a limited company in March 1914. At the time of the outbreak of World War I, the Sopwith Aviation Company had become one of the leading early British aircraft manufacturers, supplying aircraft to both the British War Office and the Admiralty (Edgerton, 1991; Driver, 1997). During the war, the youthful Sopwith transformed his small company into a mass producer of aircraft, and between August 1914 and November 1918, more than 18,000 Sopwith aircraft, of 32 different types, were designed and built for the Allied air forces. In doing so the company ended up employing 5,000 people, all managed by a man who was still only in his late 20s and early 30s (see the picture of his wartime factory in Figure 7.1).

Like a number of early aviation enthusiasts, Sopwith wasn't prepared to leave the industry in the early 1920s when there was a slump in demand, but instead held

Figure 7.1 The inside of the Sopwith factory during World War I, Richmond, Surry

on as best he could. When the war ended Sopwith tried to maintain the employment of as many of its workers as possible by diversifying into the manufacture of motorcar bodies, motorcycles and even aluminium saucepans. This strategy proved to be unsuccessful and his company eventually failed, and in September 1920, the company was put into voluntary liquidation, partly to avoid paying a war-profit tax bill from the government. "It wasn't a question of winding down. It was turning off the tap. We had scores of sub-contractors . . . and suddenly overnight, no one wanted anymore aeroplanes" (Bramson, 1990, p. 94).

At this stage, Sopwith decided to try to stay in aviation, and only two months after his company folded, he founded the H. G. Hawker Engineering Company Ltd., with himself as chairman of the board, Fred Sigrist as chief engineer and Harry Hawker as designer/test pilot. Finance for the new company was limited and mainly drawn from private investors, including Sopwith himself (*Pall Mall Gazette* 19 November 1920, p. 11; Robertson, 1970, p. 145). The new company was named after Hawker, who at the time had a national reputation as a flyer, but he was killed soon after in an air crash.

The new company's survival in the 1920s was by no means a certainty and it was to survive because it was able to develop a succession of new, innovative aircraft which were sold to the British Government. The breakthrough came in the early 1930s when increased sales to the British Air Ministry led to an improvement in its fortunes.

In June 1928 at Brooklands, the first flight of the outstanding Hawker Hart, a two-seat day bomber, took place, which led to increased sales. This aircraft was designed by Sydney Camm, who had joined the Hawker Company in 1923, and its production helped to transform the company. During the next ten years,

3,036 Harts, including seven variants of the design, were built and these eventually formed a substantial part of the Royal Air Force. The sales from these aircraft generated the profits that Sopwith was able to use to expand the company in the mid-1930s and turn what had been a small experimental company into a large industrial conglomerate. Most manufacturers at this time struggled to cope with the economic downturn of the Great Depression, but instead, the Hawker Company was able in the three years up to 1933, to generate the profits of a yearly average of £137,773 (and turnover reaching £1 million in the year ended March 1935).

The successful sale of the Hart line of aircraft meant that in 1933, it was possible to float a new public company, Hawker Aircraft Ltd., with an issued capital of £400,000 preference shares and £400,000 ordinary shares. A part of this finance was used in 1934 to buy up control of Gloster Aircraft. In January 1935, Sopwith undertook his next step and increased issued capital by £2,000,000 which was used in July to form Hawker Siddeley, which brought the firm and the Armstrong Siddeley Development company together (including Armstrong Whitworth and Avro; *Flight & Aircraft Engineer* 18 July 1935, p. 90). Sopwith was chairman of the board of the new company (the other directors were F.S. Spriggs [managing director], P.E. Hill and F. Sigrist). This new firm was far and away the largest in aviation manufacturing in the United Kingdom at this time, employing around 4,000 people in aircraft manufacture, along with more in aero engine manufacture. An additional share issue was undertaken in 1937 (£2,000,000 preference shares) to raise capital for rearmament production (*Flight & Aircraft Engineer* 18 November 1937, p. 11).

From July 1935 onwards, with acumen and skill, Sopwith began to weld together a major portion of the British aircraft industry into the Hawker Siddeley Group – combining Armstrong Whitworth, Avro, Gloster and Hawker aircraft companies with the Armstrong Siddeley Aero Engine and Motorcar Company and Air Service Training (see Figure 7.2). This huge enterprise mass-produced aircraft during World War II, delivering more than 40,000 aircraft. The company survived after the war was over, and in 1959, the de Havilland Aircraft Company was taken over, along with Blackburn and General Aircraft Ltd. in 1963. Through all of this, Sopwith remained steadfastly in charge as of the group as chairman of the board, skilfully delegating his responsibilities until in 1963, at the age of 75, he retired as chairman but remained on the board until his ninetieth birthday.

A lack of formal qualifications was no hindrance to the development of Sopwith's career. It could be argued that he adopted a planned approach to his career, exploiting opportunities as they arose – in the early days winning aircraft races and using the prize money to set up a flying school and his own company and later on using investors and money from aircraft sales to build and expand his empire under the Hawker Siddeley Group umbrella. This industrial empire encompassed airlines, aero engines and pilot training, effectively allowing him to dominate the future development of the industry in the same way that the Wrights and Curtiss had done earlier in the industry's development through patent control. His success,

Figure 7.2 Thomas Sopwith, ca. 1910

therefore, bred more success. He hung on to his power right up until the end of his life, staying in his field, developing and refining his knowledge as he went.

In a similar fashion to many of the other early pioneers/entrepreneurs, he retained a passion for adventurous sports throughout his life, even after he gave up flying. He retained an interest in outdoor sports, such as fishing, shooting and boating. A number of these were speed associated sports, and in 1913, he set up a world speed record for powerboats of 48 knots, and between 1928 and 1930, with 75 first prizes, he became the leading British 12-metre yachtsman. Along with his early aviation prizes in the United States, he later returned to that country with his J-class sloop, *Endeavour*, and came close to winning the America's Cup for the United Kingdom in 1934. In 1937, he tried again with *Endeavour II*, but lost more decisively to a much better boat (Robertson, 1970; Bramson, 1990). Despite having no more business education than any of the other aviators/entrepreneurs, he managed to learn through experience to manage his growing industrial empire, while at the same time employ gifted designers and builders of aircraft. He continued on working in the industry until his eventual retirement in 1980, perhaps the very last of the original aviation pioneers to be left in the industry.

7.4 Conclusion

The aviation industry in the United Kingdom tended to be built up by young aviation pioneers who entered the industry as part of their own personal love of flying. It was rare to see any establishment businessmen enter the industry, one exception being George Thomas Holt. Thomas Holt used his own access to finance, considerable managerial abilities and canniness to employ a gifted designer in the form of Geoffrey de Havilland to build up the United Kingdom's largest aircraft manufacturing concern in World War I.

Even established businessmen like Thomas Holt, however, struggled to maintain their businesses in the early 1920s slump in demand for aircraft. The transition from wartime military aircraft to peacetime commercial flying proved to be a difficult one, and a range of technological changes had to take place. Thomas Holt, however, did had the necessary enthusiasm before the war to establish a viable supplier of aircraft. In doing so he was able to combine the employment of gifted, innovative designers like Geoffrey de Havilland with a lifetime of experience in managing enterprises. Thomas Holt, unlike a lot of the United Kingdom's early founders of aviation companies, sold out of the industry at this time, while still retaining his support of the industry.

Thomas Sopwith was similar in that he was able to gradually acquire enough managerial experience to run a considerable industrial empire, while at the same time employed creative design teams. The creation of the Hawker Hart by the Hawker design team effectively changed the course of Sopwith's company by creating enough income from military sales to float the company and then use the funds raised to take over other companies. Sopwith then stayed associated with the industry long after the other pioneer aviators had retired. Sopwith stayed on the board of directors of Hawker Siddeley until he finally retired at the age of 90 in 1980. Still a public figure in the United Kingdom in the 1970s, he embodied for many the very epitome of the aviation pioneers/entrepreneurs.

References

Bramson, A. E. (1990). *Pure Luck: Authorized Biography of Sir Thomas Sopwith 1888–1989*. London: Patrick Stephen Ltd.

de Havilland, G. (1961). *Sky Fever: The Autobiography of Sir Geoffrey de Havilland*. London: Hamish Hamilton.

Driver, H. (1997). *The Birth of Military Aviation: Britain 1903–1914* Vol. 4. Woodbridge, UK: Boydell & Brewer Ltd.

Edgerton, D. (1991). *England and the Aeroplane: An Essay on a Militant and Technological Nation*. Basingstoke, UK: Palgrave Macmillan.

Holt Thomas, G. (1920). *Aerial Transport*. London: Hodder & Stoughton.

Robertson, B. (1970). *Sopwith: The Man and his Aircraft*. Letchworth, UK: Sidney Press.

8 From domestic aviation to military

Geoffrey de Havilland and Nevil Shute Norway

8.1 Introduction

World War I was to create considerable demand for military aircraft, which greatly boosted the prospects of the industry. Demand slumped after the end of the war and the surviving companies struggled to sell aircraft. It was only in the 1930s that the commercial prospects of air travel began to become really viable and aircraft manufacturers struggled during the 1920s. A few manufacturers in the United Kingdom and the United States were able to scratch out a living making commercial aircraft in the 1920s and 1930s, and many of the technological developments of the era were related to their efforts.

The purpose of this chapter is to look at two of the aviation entrepreneurs that were able to make a success of the construction of commercial aircraft in this period. Of these two, Geoffrey de Havilland is the best known as an aircraft designer, with many of this aircraft becoming very well known. The other, Nevil Shute Norway, was less well known as an aviation entrepreneur, but became one of the world's best-selling novelists in the 1940s and 1950s and wrote a great deal about his experiences in the aviation industry. Although in this chapter the concentration is on British manufacturers of commercial aircraft, the producers of commercial aircraft in the United States in this period are looked at in a later chapter.

8.2 Geoffrey de Havilland

Aviation was an industry that contained many interesting individuals; however, arguably the most fascinating was Geoffrey de Havilland (1882–1965). De Havilland was born in 1882 in Terriers Wycombe, Buckinghamshire, the second of three sons of the Revered Charles de Havilland and his wife, Alice Jeannette de Havilland. De Havilland was educated at the Nuneaton Grammar School, St Edward's School, Oxford, and after that at the Crystal Palace School of Engineering (in mechanical engineering). After his schooling was completed, he served an apprenticeship with Willans & Robinson, manufacturers of steam engines, until 1905 when he left the company to become a draughtsman with the Wolseley Tool & Motor Car Company in Birmingham. After that, he worked for two years in the design office of the Motor Omnibus Construction Company in Walthamstow. In a similar fashion to a number of other early aviation entrepreneurs, de Havilland had difficulty as a young man settling into steady employment, and it wasn't until he turned began to work in aviation that he found his true calling in life (Ramsden, 2015; Simons, 2017).

On seeing the flights of Wilbur Wright in 1908, de Havilland became inspired with the thoughts of flight and was able to borrow money from his maternal grandfather (an Oxford businessman) and, working with a mechanic by the name of Frank Hearle (who in 1914 married one of his sisters), designed his own aircraft and engine (Ramsden, 2015). De Havilland went onto design further aircraft and in 1910 he began to fly them. He married in 1909, and his wife, Louise, shared his interest in flying, accompanying him on many of his early flights and using a hand-turned sewing machine to make fabric coverings for some of his early aircraft. They had three sons.

Late in 1910, de Havilland went to work for the Royal Aircraft Factory at Farnborough and in 1912, flying one of his own machines, he set a new British altitude record of 10,500 feet. In the following years, he designed an important range of aircraft until January 1914, when he was appointed to be the inspector of aircraft in the aeronautical inspection directorate. He, however, always preferred working on aircraft designs, and so in May 1914, he left the Royal Aircraft Factory and became the chief designer for George Holt Thomas' Airco. During the war years, he designed and flew eight military aircraft, five of which went into war service and large-scale production (Ramsden, 2015).

As was described in an earlier chapter, the post-war slump in demand for aircraft meant that Airco closed and its assets were sold. De Havilland was keen to continue designing and test piloting new aircraft, and so he raised £20,000 and used it to start-up the de Havilland Aircraft Company with a staff of 50 (the company had a nominal capital of £50,000; *Flight & Aircraft Engineer* 7 October 1929, 1070). Prominent amongst the capital raised were sums from George Holt Thomas (£10,000), de Havilland (£3,000) and Charles Walker (£300) (Sharp, 1960, p. 71; *Flight & Aircraft Engineer* 26 September 1940, 254). Although he founded the firm and it was named after him, he chose not to become the chairman of the board or managing director, begrudging any time not spent on technical design and testing work (Ramsden, 2015). He found financial responsibilities tedious:

> It has always seemed to me that the people carrying out the work of administration and finance have the hardest, most uninteresting and thankless jobs. . . . I do know that the design and production of good aeroplanes has always been to me infinitely more important and rewarding than just making money.
>
> (de Havilland, 1961, p. 144)

At the founding of the company, the directors included A. E. Turner (chairman), de Havilland (technical director and chief designer) and Charles Walker (chief engineer). Later in 1924, a major investor in the company, Alan S. Butler, became the chairman of the board and remained so until 1950. De Havilland's brother-in-law, Frank Hearle, became the company's works manager, F.E.N. Ste Barbe the business manager and W.E. Nixon the company secretary. Hearle subsequently became a director in 1922, general manager and then managing director in 1938 (see Figures 8.1 and 8.2) (Simons, 2017).

Instead of taking one of the senior management roles in the company, de Havilland instead became and remained the head of the company's design department. This meant that in the 1920s and 1930s, he was able to devote himself to

Figure 8.1 De Havilland DH88 Comet at Hatfield in 1935

Figure 8.2 De Havilland entrance gate at Hatfield in the 1950s

designing aircraft and attempting to overcome the problems associated with the development of civil aviation. This he began to do in 1925 when he pioneered the light aircraft known as the Moth. Over 7,000 different types of Moths were built in the United Kingdom in the next seven years, with many overseas sales being made. The first flight of the Tiger Moth followed in 1931, which became one of the most-produced planes ever (Ramsden, 2015). De Havilland also conducted design on small airliners, and the de Havilland Aircraft Company manufactured 862 airliners in the United Kingdom in the 1930s, of the Dragon, Express and Dragon Rapide types (Clifford, 2015).

The role of de Havilland was crucial to the success of the company, and for it to be transformed from being a small design firm into a mass producer of aircraft. To enable mass production to take place, a successful public issue of shares in 1928 was undertaken and the company's capital raised from £100,000 to £400,000. At the same time employment rose from 400 in 1928 to 1,560 in 1929. In the second half of the 1930s further expansion occurred with the Tiger Moth, purchased by the British Government in large numbers for use in training aircrew for the RAF, and large numbers of short-haul passenger aircraft were sold.

The main contribution made by the de Havilland Company during World War II was, along with the training aircraft, the development of the Mosquito, which was perhaps the most versatile warplane ever built (Clifford, 2015). The company's employment levels rose from 5,191 in 1937 to 38,311 in 1944; aircraft production increased from 200 in 1937 to 2,327 in 1941, and the company's turnover rose from £1.4 million in 1936 to £25 million in 1945 (Simons, 2017). When the war ended, de Havilland returned to civil markets and began producing the Dove and Heron types of aircraft, but also developed the jet-powered Vampire fighter. This was followed by the development of rocket engines in 1947 and guided weapons in 1951. In 1949, the de Havilland Comet airliner first flew and pioneered the civilian jet airliner industry. Twenty-one Comets were built by the company up until 1954 when production was suspended after metal fatigue caused two machines to crash with heavy loss of life.

De Havilland himself was not directly involved in design work after 1945, and he retired from active involvement in the company in 1955, afterwards remaining as president of the company (see Figures 8.2 and 8.3). He suffered personal tragedy from his involvement in the industry; his eldest son, Geoffrey de Havilland, was killed while flying an experimental DH jet fighter near the speed of sound. Earlier his youngest son, John, had been killed in an air collision while acting as a test pilot. De Havilland's wife, Louise, who had shared much her husband's early aviation work, suffered a nervous breakdown after these deaths and died in 1949. Despite these crises, de Havilland remained committed to aviation and the pursuit of higher order learning to deliver the benefits that he believed aviation offered to society.

De Havilland career appears to mirror an entrepreneurial effectation exploratory approach (Sarasvathy, 2009) in that he focused on the pursuit of the aviation vision, acquiring knowledge and companies with expertise in that pursuit. His passion for flying took him through a number of jobs and positions with frequent

Figure 8.3 De Havilland DH104 Dove being built at Hatfield 1950

changes in jobs for variety and independence. His focus remained on design and experimentation; he was happy to work in different jobs and projects if it provided him with opportunities to pursue his interests. His passion, however, led to great family loss with the death of his two sons. His career suggests he was able to recognise and act on opportunities throughout his life. De Havilland was an enthusiastic natural historian and photographer, and insisted on remaining a test pilot while working at the company, which he did well into his 40s. He continued recreational flying until he turned 70 years old. As a design team leader, he was deeply admired by his assistants and those working for the firm, and was to a large degree the company's public face (de Havilland, 1961; Sharp, 1960).

In some ways, the way in which de Havilland created his firm was similar to what A.V. Roe did in his early years, in that he left much of the management of the factories and finances to his brother-in-law and to others associated with the firm.

This was not an uncommon approach; Herbert Royce, for instance, as a manufacturer of motorcars and aeroplane engines, took a somewhat similar approach. Even though he was not formally in charge of the firm, de Havilland played a critical role in its success. This he did both as an active member of the board of directors and as the head of the firm's design department. To a large degree, the company and its later success depended on the performance of de Havilland and his design team. It was only their success at developing popular aircraft that enabled the company to prosper. The development of the Moth in the late 1920s enabled the company to convert from being a small experimental company into a large-scale manufacturer. In the 1930s, de Havilland's development of the Tiger Moth and commercial airliners pushed this process further.

As a young man de Havilland had the usual pioneer love of speed sports such as motorcycle racing, and raised most of his initial capital from private sources. Later, success with sales of commercial aircraft enabled his company to raise further capital and change to mass production. De Havilland, like a number of other British aviation pioneers, stayed in the industry into the 1950s, maintaining the independence of his company, which finally merged into Hawker Siddeley not long after his retirement. The merging of his company into another leading company reflected the trend of more powerful survivors to subsume those with expertise and brand names after the founders had retired (Clifford, 2015).

8.3 Nevil Shute Norway

One who was influenced by de Havilland was the younger Nevil Shute Norway (1899–1960). Shute Norway was generally known as Nevil S. Norway in the industry but was also a best-selling novelist who wrote under the name of Nevil Shute. Amongst his better-known novels were *A Town like Alice*, *No Highway* and *On the Beach*. As well as being an enthusiastic flyer he was a trained aeronautical engineer who later in life often provide aviation themes as the backdrop to his novels (Smith, 1976; Thorn, 2017).

Unlike the earlier pioneers, Shute Norway attended university and graduated in 1922, aged 23 years, with a third-class degree in engineering science from Oxford University. Just as important as his formal education was his use of initial contacts provided by the Professor of Engineering at Oxford to source and undertake summer internships with the aviation companies Airco and de Havilland (in 1920 and1921) (Shute, 1954, p. 27). In both of these internships, he worked for Geoffrey de Havilland in design offices, as well as for Charles Walker. In doing so Shute Norway was able to balance his mainly theoretical studies at Oxford University, and he later admitted he found the internships far more interesting than his study.

At the time when Shute Norway first met and worked for de Havilland, the latter was an experienced aircraft designer and builder of aircraft. At first, de Havilland worked as the chief engineer at Airco, where he worked on wind tunnel research, after which he bought out some of Airco's aviation assets and then made use of them to create his own firm. Shute Norway then spent the second summer internship at the new firm and when he graduated went to work full time for the company as a calculator (stress engineer). This meant that he undertook calculations on such things as fuselage structures, propellers and wings, making use of a combination of the mathematics learnt at Oxford with the practical experience he had gained in the internships.

In this work, Shute Norway saw first-hand de Havilland's conversion of what was left of Airco into a much smaller, innovative and experimental firm (Bamforth and Abbott, 2017). Shute Norway, therefore, received a fairly comprehensive education in engineering, aviation and even in the challenges faced by entrepreneurial start-ups in the aircraft manufacturing industry. The connections and experience Shute Norway gained underpinned the development of his future entrepreneurial career. This background also supports past research on entrepreneurs, which emphasises that many have experience working in organisations before setting up their own (Bamforth and Abbott, 2017).

Despite the opportunities, by 1924, Shute Norway had become dissatisfied with what he saw as being the lack of opportunity for advancement at de Havilland. He, therefore, left de Havilland and went to work for the Airship Guarantee Company; a subsidiary set up by Vickers, especially to build airships. At this company, he became the chief calculator on the R100 airship, under the project's chief engineer, Barnes Wallis. Wallis, who was initially technical college trained (he later took extension classes at the University of London), was a brilliant scientist who later went on to develop the bouncing bomb in World War II as well as the swing-wing design in the 1950s. When Shute Norway joined the company the design work for the R100 was just starting and was part of a British Government contract to build a passenger-carrying airship in competition with a government built one (the R101) (Smith, 1976). At a very early age, therefore, Shute Norway was given considerable responsibilities. In 1929, at the age of 30 years, he was promoted to deputy chief engineer, spending some time as the acting chief engineer when Wallis left in 1930 to join the Vickers aircraft factory. The R100 then made its maiden voyage to Canada and back. Shute Norway's move to the Airship Guarantee Company meant that he enjoyed rapid promotion and was to hold positions of considerable responsibility at a relatively young age. It also provided him with the experience of seeing two companies established in the aviation industry, providing him with some knowledge of start-ups.

A future building airships was not to be, as the fatal crash of the rival R101 airship in 1930 led to the British Government cancelling the airship building programme. Vickers, in turn, closed its airship subsidiary and Shute Norway found himself looking for work. Shute Norway decoded at this point to team up with talented de Havilland trained designer A. Hessell Tiltman to found their own firm. Tiltman had worked with Shute Norway on the R100 and together they created the firm known as Airspeed Ltd. The ending of the airship programme, therefore, was to prove an important tipping point in his career, allowing him to found his own organisation at the age of only 31 years old.

Shute Norway and Tiltman's capabilities were complementary, with Tiltman having greater design experience and Shute Norway knowledge and skill as a stress engineer. Shute also had the ability to network in the industry (Anderson, 2011). Airspeed's first board meeting took place at the end of April 1931 when Shute Norway was 32 years old. In founding his company at this time, Shute Norway had a strong preference for independence and autonomy over security (Anderson, 2011). This desire for autonomy was not a unique feature of the British aviation industry. At the time, the industry had a number of famous people who relished the idea of designing and building aircraft of their own, such as Thomas Sopwith and Geoffrey de Havilland. These types of people are in contrast with men like Barnes Wallis, Sydney Camm (the designer of the Hawker Hurricane and other iconic aircraft) and R.J. Mitchell (the designer of the Supermarine Spitfire), who all lacked the same degree of entrepreneurial spirit, but instead spent their lives working for established companies.

The original intention of the firm created by Shute Norway and Tiltman was to build small aircraft and gliders for private flying recreational markets (Shute, 1954, p. 104). This market stagnated during the Depression and Shute Norway and Tiltman were forced to change their focus to the designing and building of commercial passenger aircraft (Anderson, 2011, p. 81).

The company's production was at first based in York, but later moved to Portsmouth when larger premises were needed to build the commercial passenger aircraft. At first, the AS.1 Tern glider was built for promotional purposes, and this was followed by the AS.4 Ferry ten passenger, three-engine biplane and then the Airspeed AS.5 Courier, a six-seat, single-engine, fast, light aircraft, which first flew in 1933 (Middleton, 1982; Taylor, 1970). Movement towards the building of passenger aircraft went further with the development of the Airspeed AS.6 Envoy, a twin-engine, light passenger aircraft, which proved to be fairly popular; over 50 were built and sold (Shute, 1954, p. 140). New investment was used to build the new factory at Portsmouth, and in 1934, the company became Airspeed (1934) Limited, after the Tyneside shipbuilder Swan Hunter & Wigham Richardson Limited purchased a controlling interest in the company (Middleton, 1982).

The sale of Envoys continued through the 1930s, both in the United Kingdom and overseas, and in 1936 the company joined the Society of British Aircraft Contractors (Middleton, 1982). In 1936, in response to an Air Ministry Specification issued in 1936 for a twin-engine training aircraft, Airspeed adapted the Envoy and developed it into the A10 Oxford (Taylor, 1970). The development and sale of the Envoys and Oxfords made the firm a substantial and profitable one. The Oxford went on to become the standard advanced multi-engine trainer for the RAF and British Commonwealth in World War II, with over 8,500 being built (Anderson, 2011).

Shute Norway by this stage began having trouble working with the board of the firm, as he wished to focus on designing and experimenting with new types, while the board wanted to concentrate on mass production of the Oxfords. In April 1938, aged 39 years old, he resigned as the company's managing director. After he left the company it continued, and in June 1940, the de Havilland Aircraft

Company purchased Swan, Hunter and Wigham Richardson, Ltd.'s holding of Airspeed ordinary shares. Later in 1951, Airspeed was merged with de Havilland.

When World War II broke out, Shute Norway joined the Royal Naval Volunteer Reserve as a sub-lieutenant and worked with his former boss at Airspeed Guarantee (Dennistoun Burney) in designing and creating naval engineering projects. Towards the end of the war, he had become the head of engineering in the Directorate of Miscellaneous Weapons Development, responsible for developing new equipment to help the war effort. After the end of the war, he retained an adventurous nature, and in 1948, he flew his own aeroplane to Australia and back. On returning to England, Shute Norway decided to immigrate to Australia, which he did in 1950, living there until his death in Melbourne in 1960.

Shute Norway was the type of person who looked for opportunities as best he could and at de Havilland, when he saw no opportunity for career progression, moved to the Airship Guarantee Company. Being at the start-up phase of the project, he was able to take on considerable responsibilities. He described this process as follows:

> Three consultants were employed to teach me the fundamentals of my job and carry out research into the methods. Professor Bairstow was our authority on aerodynamics, Professor Pippard on structures, and Mr. J.E. Temple was the most practical and useful of them all because he had been Chief Calculator for Wallis on the design of a former ship, the R.80, built by Vickers at the conclusion of the war. My job was to get together a staff of calculators to do the work on R.100, translating the theories of the consultants into the forces and stresses in each member of the ship and so providing the draughtsmen with the sizes for each girder and each wire.
>
> (Shute, 1954, p. 37)

The closure of the Airship Guarantee Company after the R101 disaster meant that he had to draw upon his own knowledge, experience and connections to start a new company (Bamforth and Abbott, 2017). In writing about this decision, Shute Norway said,

> After the responsibilities that I had been carrying [at the Airship Guarantee Company] I should not have relished the return to a drawing office on half the salary I had been earning; I had grown accustomed to making quick decisions and seeing them carried out. Moreover, I had very definite ideas about the design of aeroplanes.
>
> (Shute, 1954, p. 103)

> In these circumstances I decided that before seeking another job I would try my hand at starting a small aeroplane manufacturing company of my own. I was well placed to do this. . . . I went into conference with Tiltman and found that he was game to try it with me. We had a nucleus of drawing office

> staff and foremen with good aeroplane experience who had confidence in us and who wanted to go on working with us; as a technical unit to build aeroplanes we had a good deal on our side.
>
> (Shute, 1954, p. 104)

As Airspeed grew the creation aspect of the business was replaced by bureaucratic management; which Shute Norway did not especially take to. As he described later,

> Ahead of the managing director of Airspeed Ltd, stretched an unknown number of years to be spent in restraining men from spending too much time in the lavatories in order that the aeroplanes might cost the taxpayer less.
>
> (Shute, 1954, p. 167)

He made the distinction between industry leaders on the basis of whether they were entrepreneurs or managers, or as he referred to them, as 'starters or runners'. He saw himself as a starter who enjoyed the excitement of innovation, design, creative engineering and entrepreneurial activity, but who was ill-suited to the more mundane types of large-scale corporate management work. As he became disillusioned with the latter type of work, he started to reconnect to his other passion, which was writing novels. He had begun writing novels as a hobby in the mid-1920s and took it up more seriously after he left Airspeed.

He was also able to make full use of his education and experience. During the Airco/de Havilland transition, he saw how de Havilland and Walker (the firm founders) made do with little resources around them to start and build the new firm (Bamforth and Abbott, 2017). As he later commented,

> Airco at that time was near its end as a company manufacturing aeroplanes and de Havilland and Walker were already making plans to start a new company of their own. The new company was in a tiny way of business, and they were still glad to have my unpaid work, for money was very tight. To run the lathes and milling machines power was needed, and the cost of bringing electric power to the site was probably prohibitive to the infant company. Mr. Hearle and Mr. Mitchell, his foreman engineer, got hold of an old German Benz aero engine from some scrap heap; gas was available on the site, and so they set it up to run on coal gas to drive the machinery . . . for two or three years.
>
> (Shute, 1954, p. 28)

Working with Walker and de Havilland, he was able to do 'the hack work for two great designers, listening to their sane, practical experience' (Shute, 1954, p. 28). He also 'had a magnificent chance to get a knowledge of all sides of the business, and I think [he] took advantage of it . . . as the new projects came to life upon our graphs and columns of figures de Havilland and Walker still spent long periods cogitating upon our drawing boards, to [Shute's] immense benefit" (Shute, 1954, p. 28).

At the Airship Guarantee, he was again exposed to the pursuit of a vision in a totally new field with limited resources under brilliant mentors. As the airship neared completion, Barnes Wallis looked to other opportunities, allowing Shute Norway to rise to acting chief engineer within five years (Shute, 1954, p. 58).

At Airspeed he was able to put into practice what he had learnt at de Havilland and the Airship Guarantee Company. Strategic vision and determination were needed to battle through the early growth stages. As Shute Norway noted,

> There was no time to be lost, for our days were numbered till the capital ran out. . . . It was essential to get publicity and results with this machine [their first production and one of the first sailplanes developed in the United Kingdom], because our capital was fast running out.
>
> (Shute, 1954, pp. 115–116)

In running their firm both Shute Norway and Tiltman had definite views that they could not simply emulate what larger, more established firms were doing. Instead, they had to be innovative to survive. This drive for innovation suited their natural creative inclinations and created a core vision for the firm. Their strategic approach, therefore, was to design and build high-performance, innovative aircraft. In doing so, they were flexible in their response to the changing markets demands and switched from the manufacture of small recreational aircraft to commercial passenger aircraft and, finally, to military training aircraft.

This innovation meant they made important contributions to the advancement of the British aviation industry. In creating the Courier, they built the first British manufactured aircraft to have a retractable undercarriage and also used this aircraft to make important contributors to the development of inflight refuelling (Middleton, 1982). In 1933, Shute Norway was made a Fellow of the Royal Aeronautical Society, mainly because of his work on the development of the Courier's hydraulic retractable undercarriage but also for his work on the R100.

When Shute Norway left Airspeed, he did so at a time when he realised that the firm that he had helped to create was not what he had originally envisaged. As he later commented,

> In April 1938 my Board decided to attempt to quell the disputes that were plaguing the company by getting rid of me, and in this they were probably quite right. . . . In Airspeed the time for the starters was over and it was now for the runners to take over the company. I was a starter and useless as a runner; there was nothing now for me to start, and I was not unwilling to go after the first shock to my pride.
>
> (Shute, 1954, p. 169)

Instead, he went onto a rather different entrepreneurial occupation, that of a novel writer.

> I have always liked to do two jobs at the same time; one helps you to rest from the other and the fact that in the evenings my mind was fully occupied

> upon the novel gave me a clearer view of the airship problems next morning, I think, than some of my colleagues could achieve.
>
> (Shute, 1954, p. 42)

He completed his first novel in 1923 (while he was at Oxford University), but his first published novel did not appear until 1926. Except for the time that he was at Airspeed, he averaged a book a year until his death. All in all, Shute Norway wrote 24 novels and novellas. A number of his more famous novels were made into films including *Landfall*, *Lonely Road*, *Pied Piper* (1942 and 1990 as *Crossing to Freedom*), *On the Beach* (in 1959 and 2000), *No Highway*, *A Town Like Alice* (in 1956; also serialised for Australian television in 1981) and *The Far Country*.

Shute Norway's career illustrates, like others in the aviation industry, how he drew upon his own education, skills and networks to create a firm that was innovative and creative. His education and background were a particularly comprehensive one, combining an academic education with applied training and exposure to both leading aircraft designers and entrepreneurs at a very young age. He was to benefit from exposure to two start-up firms (de Havilland and the Airship Guarantee Company) before starting up his own. Like others (the Wright brothers, Curtiss, etc.), Shute Norway also contributed significantly to the development of innovative products that fundamentally changed the industry; for example, a version of Shute Norway's hydraulic retractable undercarriage is still used in aircraft today.

The most important aspect of his entrepreneurial vision was that although he had clear ideas of the firm that he wished to set up and the types of aircraft he wanted to produce, he and Tiltman quickly changed their approach in changing circumstances. This was a result of the nature of the industry, which at the time was going through a period of both economic slump and great technological change. These changes in technology meant that there were increases in economies of scale and changes in demand for aircraft away from private markets towards commercial passenger and military aircraft. It made it difficult for the type of firm envisaged by Shute Norway and Tiltman to survive. Later in the post-World War II period, there were a number of small-scale manufacturers of light, recreational aircraft that prospered, but in the 1930s, it was the development of commercial passenger aircraft and later military aircraft that dominated the industry. The technical dynamism of the industry that had attracted Shute Norway to the industry and its opportunities eventually led to the creation of conditions that made it difficult for him to stay in the industry.

8.4 Conclusion

Both de Havilland and Shute Norway worked in the period between the wars when the aviation industry turned towards commercial developments as a means forward. Both had a dislike of having to deal with bureaucracy and military purchasing agents and so were keen to develop commercial aircraft as the market emerged. This market, however, took time to develop, largely due to lack of government funding. Both were able to adapt to these circumstances, with Geoffrey de Havilland, in particular, being able to design aircraft well suited to the market.

What differed between the two, however, is that de Havilland managed to maintain his position as chief designer at his company and work with the company's managers and company directors. Shute Norway found that more difficult, first taking on the role of managing director, which involved less direct work in design. This was not unusual in the aviation industry, with men like Sopwith, for instance, moving in a similar direction, but Shute Norway found the work less inspiring and finally ended up leaving the industry after a dispute with his company's board of directors.

The careers of both men illustrate the tensions that existed in aviation manufacture in its earliest days, in that innovative design teams had to sit with corporate structures, which over time got larger in scale. Dealing with these tensions was an important problem which de Havilland was able to overcome, Shute Norway less so.

References

Anderson, J. (2011). *Parallel Motion: A Biography of Nevil Shute Norway*. New York: Kerhonkson.

Bamforth, C. J. and Abbott, M. (2017). Modelling the behaviour of an entrepreneur: The case of Nevil Shute and the British aviation industry. *International Journal of Innovation*, 9(3), pp. 3–26.

Clifford, J. (2015). *De Havilland and Hatfield 1910–1935*. Toadsmoor Road, UK: Fonthill Media.

de Havilland, G. (1961). *Sky Fever: The Autobiography of Sir Geoffrey de Havilland*. London: Hamish Hamilton.

Middleton, D. (1982). *Airspeed: The Company and its Aeroplanes*. Sudbury, UK: Terance Dalton Ltd.

Ramsden, J. M. (2015). *Sir Geoffrey de Havilland: A Life of Innovation*. Welwyn, UK: Hylands Publishing.

Sharp, C. M. (1960). *DH: A History of de Havilland*. London: Faber.

Shute, N. (1954). *Slide Rule: The Autobiography of an Engineer*. London: Heineman.

Simons, G. M. (2017). *De Havilland Enterprises: A History*. South Yorkshire: Pen and Sword.

Smith, J. (1976). *Nevil Shute*. Boston, MA: Twayne Publishers.

Taylor, H. A. (1970). *Airspeed Aircraft Since 1931*. London: Putnam Publishing.

Thorn, R. (2017). *Shute: The Engineer who Became the Prince of Storytellers*. Kibworth Beauchamp UK: Troubador Publishing.

9 The engine makers

Henry Royce and Frederick Brant Rentschler

9.1 Introduction

One of the main reasons why heavier-than-air flight was made possible was the development of more high-powered, petrol combustion engines. It was these that gave the pioneer aviators the power they needed to get their aircraft into the sky. The advances made by the companies that manufactured these engines, therefore, was an important part of the overall development of the industry after 1903.

Compared to the manufacture of airframes, however, the production of aero engines was a more complex, capital intensive process and so there were fewer companies that were engaged in this type of activity. Perhaps as well it was less attractive to the large number of aviation pioneers who felt that flying was a great deal more exciting than developing engines.

At the beginning of the twentieth century, aero engines were fairly simple, low-powered machines, generally designed and built for a specific aircraft. Soon, however, aero engines began to be manufactured in quantity and in a number of cases by several manufacturers, each located in a different country, who were licenced by the designer/initial manufacturer of the engine (Heron, 1961). In the United States, especially during World War I, motorcar manufacturers dominated the production of aero engines; however, firms specialising in their production were created in the 1920s. The earliest aero engines were stationary in type, either in line or radial in style, and were gradually succeeded by the popular rotary engine (Heron, 1961).

In the United Kingdom, the Rolls-Royce Eagle, which entered production during World War I, and its successor, the Falcon, marked the beginning of a famous line of engines that later resulted in the famous Merlin and Griffon engines of the World War II era (Harvey-Bailey, 1982). The most important of all engines produced by the Allied nations in World War I, and perhaps the United States' greatest contribution to wartime aviation, was the Liberty engine. In May 1917, Jesse G. Vincent of the Packard Motor Car Company and E.J. Hall of the Hall-Scott Motor Car Company stayed at a hotel in Washington, DC, and for nearly a week and worked on the design of the 8- and 12-cylinder Liberty engines, which they intended to be mass-produced (Heron, 1961). The Liberty became the standard aero engine of World War I and was used most often in powering the DH-4, the only United States made aircraft to go into combat on the Western Front (Heron, 1961).

The motorcar industry, which had made use of combustion engines, was a common source in most countries of expertise, which was used to produce aero engines. Two of the most well-known entrepreneurs that specialised in the manufacture of aero engines were Henry Royce in the United Kingdom and Frederick Brant Rentschler in the United States. In this chapter, these two entrepreneurs are examined in order to see how central the manufacture of engines was to the development of the industry.

9.2 Henry Royce

In the United Kingdom, the most well-known aero engine manufacturer was Henry Royce (1863–1933). He was an engineer and motorcar designer by occupation, and was born on 27 March 1863 at Alwalton, near Peterborough, the youngest of two boys and three girls of James Royce, a flour miller, and his wife. When Royce was four, his father took his two sons to London, intent upon running a metropolitan flour mill. This venture was not successful and when financial trouble arose the young Royce became a W.H. Smith newspaper boy, first at Clapham and then at Bishopsgate. Royce's father died when he was nine, and his schooling afterwards lapsed; for a time, Royce worked as a post office telegraph delivery boy.

After Royce turned 14 an aunt paid for him to be apprenticed to the Great Northern Railway Company's locomotive works in Peterborough. He worked there for three years, until his aunt's funds ran out, but fortunately for Royce, his enthusiasm and skill as a workman meant that he had been able to acquire the knowledge of the trade in that time. The work stimulated Royce's fascination with engineering, which was to be a feature of his work through all of his life. Royce was forced to tramp the roads of England looking for work and eventually found it in a machine tool firm in Leeds. Later, he decided to move to London, and there he became a tester with the London Electric Light and Power Company, which at the time was installing electric arc and incandescent lighting in London's streets. At this time, he was able to extend his schooling and attended technical classes at Quintin Hogg's Polytechnic Day School and evening classes at the City and Guilds Institute. Later still he later became the chief electrical engineer of a subsidiary, the Lancashire Maxim and Western Electric Company, a firm established to introduce electric lighting to Liverpool (Pemberton, 1942).

Royce's early education and experience are similar to that of a number of the British aviation founders, except he did not have the support of middle-class parents. He did, however, eventually acquire a technical college education along with experience in heavy engineering (Pemberton, 1942). He also seems to have exhibited the general restless in employment that many of the others did, changing from one job to another in his enthusiastic pursuit of engineering.

In undertaking his work in Liverpool, Royce met Ernest Albert Claremont, and in 1884 (when Royce was 21), together they founded F. H. Royce & Co. in

Manchester as a manufacturer of electric bell sets, lamp holders, switches, fuses and registering instruments. For this company, Royce designed a drum-wound armature for a dynamo and these dynamos became widely used in lighting cotton mills, factories and ships. This equipment was to gain a strong reputation for reliability and longevity, which were engineering characteristics to which Royce gave a very high priority. The partnership of Royce and Claremont was further strengthened on 16 March 1893 when Royce and Claremont married two sisters, daughters of Alfred Punt, a licenced victualler of London (Pugh, 2000).

The company continued to prosper, and in 1894, the two partners converted their firm into a limited company, Royce Ltd., with Royce focusing on the development of technical innovations, which at the time included electric cranes and motors for lock gates. Between October 1897 and February 1899, the value of their company's orders rose from £6,000 to £20,000 and the company's capital was increased to £30,000 in order to finance an expansion of the firm's productive capacity. Later, this growth of the company was to stagnate, and Royce's interest began to shift from cranes to road vehicles.

Royce first bought a motorcar in 1903, when French and German manufacturers dominated the industry, and he attempted to redesign his purchase. In autumn 1903, he announced that he intended to build his own motorcar and in the following year his first motorcar was built. Royce subsequently tried to improve its reliability, and it was at this time that he met Charles Stewart Rolls, who ran an agency for the sale of French cars in London's West End. Rolls had come to hear of Royce's efforts, and in May 1904, he travelled to see Royce's car. He was so impressed with the car that Royce had built that he negotiated to become Royce's sole agent and so began the famous partnership of Rolls and Royce (see Figure 9.1) (Harvey-Bailey, 1982).

Figure 9.1 Henry Royce

While Rolls advertised Royce's cars in a series of reliability trials and races during the years 1904 to 1906, Royce focused on the design and production of new models of engine. To begin with, Royce produced a variety of models based on four chassis of 10, 15, 20 and 30 h.p., the most successful being the four-cylinder 20 h.p. Grey Ghost (Harvey-Bailey and Evans, 1984). As sales proved to be successful, Rolls-Royce Ltd. was formed on 15 March 1906, and Royce became the firm's chief engineer and works director. As such he was the most highly paid man in the firm. At first, they experienced difficulties in raising capital; however, in December 1906, Royce began work on the design of the 40/50 h.p. Silver Ghost, which was to remain in production, substantially unchanged, from 1907 until 1925, when it finally gave way to the production of Wraiths and Phantoms.

In order to cope with the demand for the Silver Ghost, in June 1908, the motor section of the company was separated from the plant at Cooke Street, Manchester, and moved to a new location in Derby. By this time Royce was seen by his firm's board as being a brilliant designer but as a poor production engineer. When Royce became seriously ill in September 1908, he was persuaded by the board to work in a drawing office at home with the assistance of a team of draughtsmen. This became a common way for Royce to work for much of the rest of his life.

Later still, the news of Charles Rolls' death in 1910 in an air crash triggered a further breakdown in his health, and in 1911, he underwent a major operation. After his health improved, he began to live in a villa in the south of France along with a personal staff of eight who lived in adjacent premises and worked in his workshop. This unusual (for that time) separation of production from design was to last for the rest of his career, and he divided his time between his homes at St Margaret's Bay, Kent (to which he moved in 1912) or West Wittering in Sussex (to which he moved during World War I) in summer, and the villa at in the south of France at Le Canadel in the winter. Despite the irregular nature of this arrangement, Royce was able to control the use of his designs and was able to keep in close contact with the various advancements taking place in the industry.

Although before the war Charles Rolls had often pressed him to design an aero engine, Royce initially took no practical interest in the matter until after the outbreak of World War I in 1914. Indeed, after the death of his partner, Royce had a strong aversion to any participation in the industry. It was only after he saw an airship struggling to cross the channel with too weak an engine that he was convinced to modify the Silver Ghost engine for use in aircraft (Lloyd, 1978). This work on engines was important to the British war effort because at the time the country was heavily reliant on French-built engines. When World War I broke out, officials from both the War Office and the Admiralty requested that Rolls-Royce also build aero engines.

After he had investigated various types of air-cooled engine, he decided to concentrate on liquid cooling engines. He started a 12-cylinder V engine and then in early 1915 produced the 200 h.p. Eagle for the Admiralty. This engine was initially tested at 225 h.p. in March 1915, and its later versions were able to produce as much as 360 h.p. It powered, along with a number of other important aircraft, twin-engine bombers including the Handley Page 0/400 that later became a successful airliner, and the Vickers Vimy that, in 1919, became the first aircraft to be flown across the Atlantic Ocean (Harvey-Bailey and Evans, 1984).

Rolls-Royce followed the Eagle with the Falcon, the Hawk and later the Condor. Overall, in World War I, Rolls-Royce was to supply more than 60 per cent of all the British-built aero engines (Lloyd, 1978). When peace came at the end of 1918, a marked fall off in demand took place for these engines, and Rolls-Royce turned back to the production of motorcars. This situation did not last long, however, and during the mid-1920s, the aircraft builder Richard Fairey encouraged Rolls-Royce to make a renewed commitment to the production of aero engines.

During the 1920s, the firm continued to manufacture aero engines but on a far smaller scale than during the war, and at Royce's instigation, it entered the Schneider Trophy competitions in 1929 (Lloyd, 1978). At this time Royce modified the 850 h.p. Kestrel engine (which he had designed in 1925) into the 'R' engine. This engine was installed in the Supermarine S6 seaplane, designed by R.J. Mitchell, which subsequently won the Schneider Trophy in 1929. Further design work undertaken by Royce led to improvements to the engine, which enabled the United Kingdom to keep the Schneider Trophy in 1931. From this experience, as well as that gained from transatlantic competitions, Royce laid down the prototype for his next engine, the Merlin, a 12-cylinder V engine, which later powered the Spitfires, Hurricanes, Mosquitos, Lancasters and Mustangs in World War II (Harvey-Bailey and Evans, 1984). Struck by ill health, Royce died in 1933 without seeing just what a contribution his engines made to the British war effort.

A bit like de Havilland, Royce concentrated on the design and development of new products rather than on the management of the firms he was associated with. As such Royce in his career had only two directorships, at Royce Ltd. of Manchester, and Rolls-Royce Ltd. of Derby. He was, however, a member of the institutions of mechanical, electrical and aeronautical engineers and a very prominent personality in the industry. He tended to avoid direct involvement in the overall management of Rolls-Royce but instead concentrated on designing new products for the firm. In doing so, he was never the managing director or chairman of the board of his companies but instead relied on others to undertake this work. This sort of arrangement was not common but did exist at the time. It reflected the manner in which de Havilland structured his company and the way A.V. Roe initially structured his affairs.

The disruptive early start of Royce's life led to the necessity for him to pursue opportunities that would allow him to quickly become self-sufficient. Like many others at the time, he started his first company in his early 20s, which was hugely successful. While others caught the aviation bug, Royce caught the motorcar bug. The partnering of Royce's technical and design skill with Rolls' sales and administrative capability within Rolls-Royce Ltd. provided a good match. Rolls used the media to promote Royce's cars at a time when high-quality British motorcars were unheard of. The outbreak of war alerted Royce's attention to the need for high-powered, high-quality aviation engines. Like before, these were promoted through trophy competitions in 1929 and 1931 with Royce drawing upon metallurgical research at Derby to beat the competition. His membership in mechanical, electrical and aeronautical engineer institutions gave him access to prominent figures in all these industries. Although Royce's latter career was stable and structured, this was a by-product of pursuing his passions, which he was able to protect and pursue despite being part of large and influential organisations.

9.3 Frederick Brant Rentschler

In the United States, despite the early start of the aero engine manufacturing industry, it lagged behind that of France and the United Kingdom until the mid-1920s. One person who was to help overcome this gap was Frederick Brant Rentschler (1887–1956). Rentschler was an aviation engineer, aircraft engine designer, industrialist and most importantly the founder of Pratt & Whitney Aircraft. In his career, he was to create and to manufacture many revolutionary aircraft engines, including those used in the aircraft of a number of famous aviators such as Charles Lindbergh, Amelia Earhart and James Doolittle. In addition, he was also a co-founder of the United Aircraft and Transport Corporation (UTAC), the predecessor of the United Technologies Corporation (Rentschler, 1950).

He was born on 8 November 1887 in Hamilton, Ohio, to two German-Americans, George A. Rentschler and Phoebe Schwab, whose wealthy family owned the Republic Motor Car Co., which manufactured Republic cars from 1910 until 1916. In addition, the family was also principal in Hooven-Owens-Rentschler, manufacturers of steam and diesel engines, and his brother, Gordon S. Rentschler, became the chairman of National City Bank.

Rentschler was to graduate from Princeton University in 1909 and initially he worked in his family's businesses as a moulder and machinist. After the United States entered World War I in 1917, he joined the United States Army and as a First Lieutenant and later Captain, he was given the job of inspecting the Hispano-Suiza aero engines that were manufactured under French licence at the Wright-Martin plant in New Brunswick, New Jersey. The end of the war in 1918 meant the end of this contract and at the same time caused the reorganisation of Wright-Martin.

At this time Rentschler left the Army and was convinced that the future of aviation would depend upon the development of lighter-weight engines with much greater power and reliability. Against the conventional wisdom of the day, he proposed the design of an air-cooled engine. At this time, it was felt by many that the heavier, liquid-cooled engines were the way of the future.

With the reorganisation of Wright-Martin, Rentschler became the president of the Wright Aeronautical Corporation that emerged, and he pressed hard for research into his ideas in engine manufacturing. He was, however, not able to convince his board of directors of his plans, the board largely being composed of investment bankers who had little aviation knowledge. He decided to resign from the company in 1924 and, supported by old friend and Wright chief engineer George J. Mead, he developed a proposal for a high-powered, air-cooled aircraft engine for the United States Navy. In doing so, he was given encouragement by naval authorities, with Admiral William A. Moffett, in particular, promising to approve the purchase of such an engine by the Navy.

To further advance his ideas Rentschler approached the Pratt & Whitney Machine Tool Company of Hartford, Connecticut, and on 23 July 1925, the company agreed to fund its development. To do this they created the Pratt & Whitney Aircraft Company, with Rentschler and Mead taking a controlling position.

The manufacture of Pratt & Whitney Aircraft's first engine was completed on Christmas Eve 1925 and was named the Wasp by Faye Belden Rentschler, whom Frederick had married on 25 July 1921. This engine had 425 h.p. and passed its official qualification test in March 1926. The Navy made good on its promises and ordered 200 of them – a large purchase for the mid-1920s. The speed, climb performance and reliability of this engine was to revolutionise American aviation. From this original Wasp design, the Pratt & Whitney Wasp series developed over the next 20 years, leading to the development of engines approaching ten times the power of the 1925 engine (see Figure 9.2). Many iconic aircraft of the era made use of the engine including the DC3 and the B17 flying fortress.

In 1929, Rentschler was to end his association with the Pratt & Whitney Machine Tool Company, but he was allowed to keep the name Pratt & Whitney Aircraft Company, which by this time was becoming more famous than its parent. Around this time Rentschler, along with Chance M. Vought and William Boeing, formed UATC, which completed the first coast-to-coast passenger network in March of that year. This firm was not to survive in one piece, however, but was broken up in 1934 with its manufacturing interests east of the Mississippi River becoming the United Aircraft Corporation, headquartered in Hartford with Rentschler as president.

Rentschler's Pratt & Whitney competed vigorously with Wright throughout the 1920s and 1930s for the engine market, and the two came to dominate the American industry. The extreme precision needed to produce aero engines prevented other firms from entering the industry. The need for a particular aircraft to typically use a single-engine model for all of its units led to a fairly steady stream of orders for engines as long as that aircraft was used. As engines wore out quicker than most aircraft, a steady demand for engines was kept up.

As well as the military market, Rentschler's Wasp engine also came to dominate the civil market, with the aircraft builders of the 1930s designing their new commercial passenger planes and transports around the Wasp engine. In all, these engines came to power around 100 different aircraft types, including some of the fastest and most famous aircraft of the time.

The firm continued to grow, and by 1930, Pratt & Whitney dominated around 60 per cent of the aero engine market in the United States. By the early 1930s, Pratt & Whitney was also building the Twin Wasp. This engine was later to power many of the bombers, fighters and transport aircraft of the period. Later in 1938 and 1939, desperate to expand its air force, France ordered some US$85 million worth of the engine, which helped Pratt & Whitney to finance the expansion of manufacturing facilities in the lead up to the outbreak of World War II.

During the war itself, some 173,618 Twin Wasp engines were produced, more than any other engine type in the history of the industry. Late in the war,

Figure 9.2 A workman on the final assembly of a Pratt and Whitney Wasp engine, East Hartford, Connecticut

Rentschler turned to developing jet engines, and after World War II, the Pratt & Whitney Company produced the J57 jet engine. The engine was used to power the first B-52 Stratofortress in 1954 along with a number of other aircraft. Rentschler remained as president of Pratt & Whitney and United Aircraft until his death on 25 April 1956.

Rentschler's early family exposure to successful company operations in cars, steam and diesel engines gave him technical and administrative exposure to the running of successful company operations. It was, however, his World War I exposure inspecting aircraft engines that made him aware of the opportunities to enhance them. Inability to secure funding, despite his connections, did not prevent him from starting his own company with Pratt & Whitney. No doubt personal money, connections and an innovative engine design helped. At this point, he moved away from a stable career in the family business to pursue his vision of better aircraft flight. But four years later, the partnership had run its course, and he moved on to another start-up (United Aircraft) with Boeing and Vought until his death. In this position, he developed several successful jet engines.

9.4 Conclusion

Although the main motivation of men like Rentschler and Royce was not an obsession with flying, they did both exhibit a strong notion of the types of designs they wished to pursue and were prepared to leave conventional careers as engineers to pursue their ideas. It was only by founding their own companies that they were able to achieve the dreams that they had, so in a way, their entrepreneurial spirit was driven by the demand of their engineering plans. Royce even went as far as leaving the job of running the companies he founded to other associates so he could concentrate his efforts on design and development. But even Rentschler turned his back on quite lucrative and stable careers, first with the family business and then with Wright to pursue his own more risky ambitions.

References

Bastow, D. (1989). *Henry Royce-Mechanic*. Derby UK: Rolls-Royce Heritage Trust.

Harvey-Bailey, A. (1982). *Rolls-Royce: The Formative Years, 1906–1936*. Derby: Rolls-Royce Heritage Trust.

Harvey-Bailey, A. and Evans, M. (1984). *Rolls-Royce: The Pursuit of Excellence*. Paulerspury, Northamptonshire, UK: Sir Henry Royce Memorial Foundation.

Heron, S. C. (1961). *History of the Aircraft Piston Engine: A Brief Outline*. Detroit, MI: Ethyl Corporation.

Lloyd, I. (1978). *Rolls-Royce: The Growth of a Firm*. London: Palgrave Macmillan.

Pemberton, M. (1942). *The Life of Sir Henry Royce*. London: Hutchinson & Co. Ltd.

Pugh, P. (2000). *The Magic of a Name: The Rolls-Royce Story: The First 40 Years*. London: Icon Books.

Rentschler, F. B. (1950). *Saga of Pratt & Whitney Aircraft Co*., East Hartford, CT: F.B. Rentschler.

10 A new era of transport

James Smith McDonnell, Donald Willis Douglas and William Boeing

10.1 Introduction

At the same time that de Havilland and Shute Norway were developing commercial aviation in the United Kingdom, American entrepreneurs were developing this aspect of the industry on an even greater scale. In the United States, the principal post-World War I survivors of shakeout were Curtiss, Martin, Wright Aeronautical (by now mainly making aero engines) and Boeing. The Boeing Company began in 1915 when William E. Boeing and Naval Commander Conrad Westervelt entered into a partnership to develop a seaplane. After the war, it produced a fighter, which led to it developing fighter design capabilities. United States Government demand remained uncertain and relatively low, so aviation firms looked towards commercial and later export markets in the 1920s and 1930s to maintain their profitability. In the United States, commercial aircraft manufacturing surpassed that of military production throughout most of the 1930s, and Boeing, in particular, concentrated on developments in this field to make sales. In addition, entrepreneurs such as James Smith McDonnell, Jack Northrop, Allan Loughhead and Donald Douglas were important, and all of them appreciated the potential to develop faster, more efficient airliners.

With the support of venture capitalists, Loughead established in southern California the Lockheed Aircraft Corporation in 1926 and employed Jack Northrop to develop a radical new airliner design, the Lockheed Vega. This plane was a streamlined, high-speed design which used a single, unbraced wing at the bottom of the fuselage. The Vega differed considerably with the slower, high-wing Ford and Fokker trimotor airliners, which dominated civil transport in the late 1920s. The Vega was first flown in 1927, and it proved to be commercially successful. Lockheed then integrated a number of the new innovations into a new plane the company developed, the Alpha, which was first flown in March 1930. Boeing was also making changes and developed a high-speed airmail and transport plane known as the Monomail, using the same radical design approach and technologies that Lockheed did.

Donald W. Douglas, who in the early 1920s had resigned as chief engineer of the Glenn L. Martin Company, moved to Los Angeles in order to go into business for himself. Douglas also began to produce innovative designs, gaining the support of local businessmen becoming a constructor of Army utility aircraft and conventional Navy torpedo bombers. He later, most notably, produced the radical DC-1 design, which incorporated many of the revolutionary design features pioneered by Boeing and Lockheed.

At the same time, American fighter development lagged behind that of the European powers, despite the developments in commercial aircraft and associated developments in bomber aircraft. Late in the 1930s, the overseas markets developed, which encouraged the development of fighters. The emergence of a potentially profitable foreign market, more viable civil aviation, and the high-tech overlap between commercial and combat aircraft (especially bombers) had the effect of drawing new entrants into the industry. Expansion of opportunities in the 1930s led to the entry of new companies such as Bell, Northrop and Vultee.

The purpose of this chapter is to look at the American entrepreneurs who developed commercial aviation in the 1920s and 1930s – in particular, McDonnell, Douglas and Boeing. The successor companies of those established by these men still exist in the United States today and are amongst the best known in the industry.

10.2 William Boeing

William Edward Boeing (1881–1956) was born in Detroit, Michigan, to Wilhelm Böing from Germany and Marie M. Ortmann, from Austria. Born into a successful commercial family, Wilhelm Böing emigrated to the United States in 1868, initially working as a labourer. Moving to America was not popular with his father and he, therefore, received no financial support from his family. This did not hamper him, and he went on to make a fortune from mineral rights near Lake Superior and North Woods timberlands and mineral (Bauer 1990; D'Intino et al., 2008; Rogers, 1996).

At the age eight in 1890, Boeing lost his father to influenza, and his mother soon remarried. Because of his family's wealth, he attended school in Europe at Vevey, Switzerland, and later returned to the United States to attend a year of prep school in Boston. Back in the United States, he changed his name to William Boeing and attended Yale University, but left there in 1903 without graduating to go into the family's timber business.

In pursuit of this career, he moved to Seattle in 1903, and there he bought timberlands around Grays Harbor on the Pacific side of the Olympic Peninsula in Oregon. He also purchased lumber operations. These ventures were a success, which was in part due to shipping timber to the East Coast via the new Panama Canal. It was from this business that he raised the funds necessary to enter the aviation business.

While Boeing was president of Greenwood Timber Company, he experimented with boat design, and in doing so travelled to Seattle, where, in 1909, during the Alaska-Yukon-Pacific Exposition, he observed a manned flying machine for the first time. In 1914, William Boeing and his friend Conrad Westervelt paid for several aircraft rides and he decided to take lessons at the Glenn L. Martin Flying

School in Los Angeles. He also purchased one of Martin's planes (Serling, 1992, p. 2). He and Westervelt enjoyed the experience of flying so much that they then set out to build a better aircraft, which was the seaplane Boeing Model 1, which first flew in June 1916. They tried to sell it to the Navy without success, but the Boeing Model C, which they subsequently built, sold well after America's entry into the war. To carry out this production, Boeing purchased an old boat works on the Duwamish River near Seattle and used this as his first factory to build aircraft. His company became known as the Pacific Aero Products Co. but was altered when America entered World War I to the Boeing Airplane Company.

After the war Boeing concentrated on the development of commercial aircraft and was able to secure contracts to supply airmail services. From this operation evolved the successful passenger service that became United Airlines. The post-World War I collapse in aircraft demand meant that Boeing made a variety of non-aviation products (such as bedroom furniture) in line with others in the industry at the time and raised capital with new stock issues (Serling, 1992). Boeing was able to make some sales of aircraft, the company building 25 HS-2L flying boats for the Curtiss Company. At this time, Boeing found it necessary to use his own funds to cover the company's expenses and to keep it solvent. In 1921, an order from the Army Air Service to build 200 Thomas-Morse MB-3A pursuit fighter biplanes kept his company going and put it onto the road to financial success. During the early 1920s, Boeing sold planes to both the United States Army and Navy, becoming a leading aviation firm by 1927, employing about 800 employees. Boeing became the dominant supplier of naval fighters in the 1920s and 1930s, and in 1928, the Boeing Airplane and Transportation Co. went public, selling its stock to aggressive investors.

At around this time, Boeing became friends with Fred Rentschler, the president of Pratt & Whitney, whose engines from the mid-1920s onwards were used on Boeing aircraft. In 1928, Rentschler suggested merging BAT and Boeing Airplane Company with Pratt & Whitney into a holding company. Boeing agreed to this move, and on 1 February 1929, the UATC was incorporated. This new firm was a large holding company that included the engine manufacturer Pratt & Whitney, the aircraft manufacturers Boeing, the Hamilton Metalplane Company and the Chance Vought Corporation. UATC also created the subsidiary Boeing Aircraft of Canada, which began building C-204 flying boats. As part of its expansion UATC also purchased three airlines and (in July 1929) Sikorsky Aviation (which specialised in amphibian aircraft) and Stearman Aircraft of Wichita. As part of this company's holdings a transport group consisting of BAT, National Air Transport, Varney and Stout Airlines was set up, which evolved into United Air Lines.

These types of conglomerates became unpopular with the American public during the Great Depression, and in 1934, the United States Government accused William Boeing of monopolistic practices. The *Air Mail Act* was passed in the same year and it forced aircraft companies to separate flight operations from the development of new aviation producers and manufacturing. William Boeing was forced to divest himself of ownership of his holding company, UTAC, which was subsequently separated into three businesses. These were the United Aircraft

Corporation (which held the former eastern United States manufacturing, now known as United Technologies Corporation); the Boeing Airplane Company (with western United States manufacturing, which later became the Boeing Company); and the United Air Lines (flight operations).

Boeing, unhappy with this clash with the United States Government, began to withdraw from the industry, preferring instead to invest most of his time and money in property development and thoroughbred horse breeding. Between 1935 and 1944, William Boeing and his wife, Bertha, used a large section of land to the north of Seattle for subdivision, which later became the future communities of Richmond Beach, Richmond Heights, Innis Arden, Shoreview and Blue Ridge. Racially restrictive covenants on their land were imposed by the Boeings in order to enforce segregation. William Boeing died in 1956, at the age of 74.

The legacy of Boeing was one of technological vision, innovation, improving technology and becoming and remaining an industry leader. Boeing's primary interest was on innovating new aircraft designs, even if it meant a reduction in short-term profits (Collins, 2001; Collins and Porras, 1995, 1997; Mansfield, 1966; Redding and Yenne, 1983). Boeing used his personal fortune and company funds to design and build new prototypes and developed a reputation for industry leadership. Although he was one of the most successful in this fashion, he was not alone in taking this approach. Most of the pioneers of aviation invested much of their own and their families' capital into their obsession with flight. In addition, staying in the industry in the early 1920s took considerable courage and a willingness to withstand financial hardship.

As Boeing was in the second wave of entrepreneurs coming into the United States marketplace (behind the Wrights and Curtiss) he benefitted from their accumulated knowledge. Planes were available for him to purchase, fly and learn from and he was able to deliver designs responsive to United States Government needs. Initial support from the American military in the lead up to the country's entry in the war and then post-war government interest and investment in commercial flight allowed Boeing to build a dominant and initially monopolistic position in the industry – family money and connections no doubt helped this process. In this way, Boeing followed the Wrights' and Curtiss' lead in seeking to dominate the market. Boeings career was initially entrepreneurial but became more strategic and staged as his power over the market grew.

10.3 Donald Wills Douglas

The great rival to the Boeing Company in the post-World War II years was that of the Douglas Aircraft. The company's founder, Donald Wills Douglas (1892–1981) (see Figure 10.1), was an American aircraft engineer and industrialist. An aviation pioneer, he designed and built the Douglas Cloudster and later a range of commercial aircraft that helped to change the nature of world commercial aircraft. Although the Cloudster failed in its intended purpose, which was to be the first aircraft to travel non-stop across the United States, it became one of the first aircraft that had a payload greater than its own weight (Cunningham, 2010; Morrison, 1991).

Figure 10.1 Donald Douglas with some wing ribs

Douglas was born in Brooklyn, New York, in 1892, the second son of an assistant cashier at the National Park Bank. He attended the Trinity Chapel School, and after graduating from there in 1903, he enrolled in the Naval Academy in Annapolis, Maryland. In doing so he followed an evolving interest in sailing and later in flying. At the Academy, he studied (amongst other things) marine engineering, which gave him a basic knowledge to tackle some of the aeronautical problems that he was later to face. In a visit with his father to the Smithsonian Institution, he observed the aircraft engine that Charles Manley had built for Samuel Langley. He then became an early aviation enthusiast, and at the age of 16 in the fall of 1908, he witnessed the Fort Myer trials of the Wright Flyer by Wilbur and Orville Wright, which they undertook for the United States Signal Corps (Filucci, 2018, p. 22). He also observed Glenn H. Curtiss flying one of his planes at Long Island in 1909 and 1910 (Filucci, 2018, p. 23). He took up the hobby of building model aircraft, some with rubber-bands and later other motors, in his dormitory room at Annapolis. He then tested them on the grounds of the Academy. He resigned from the Academy in 1912 in order to try to follow an aeronautical engineering career.

After being turned down for employment by people such as Grover Loening and Glenn H. Curtiss, Douglas decided to enrol at MIT. In 1914, he received the bachelor of science in aeronautical engineering degree, the first person to be conferred the new degree at MIT, after completing the four-year course in only two years. Post-graduation he stayed at MIT for another year working as an assistant to Professor Jerome Hunsaker, an aeronautics expert.

Douglas joined the Connecticut Aircraft Company in 1915 and took part in the designing the DN-1, the Navy's first dirigible. In August 1915, Douglas left this company to join the Glenn L. Martin Company where at 23 years old he became the chief engineer. Shortly after joining, Glenn L. Martin Company merged with the Wright Company to form Wright-Martin, and Douglas, unhappy with this change, resigned in November 1916 to become the chief civilian aeronautical engineer of the Aviation Section of the United States Army Signal Corps. In the military, he spent his time redesigning British aircraft for American use and developing the Liberty engine.

Later on, Douglas returned to the newly reformed Glenn L. Martin Company in Cleveland, Ohio, again becoming the company's chief engineer. In his time with the company, Douglas would design the Martin MB-1 bomber. After wartime work with Glenn L. Martin in California and Ohio, Douglas returned to the West Coast and in Santa Monica, California, set up his own company. Initially, the company was known as the Davis-Douglas Aviation Company as it was largely financed with US$40,000 from his partner David Davis, a wealthy sportsman interested in aviation. This investment by Davis in Douglas' company was on the condition that Douglas design and build a single aircraft which would be used to attempt a non-stop crossing of the United States. In response to this offer, Douglas agreed, and the Davis-Douglas Company was created on 22 July 1920. After the aborted flight of the company's Cloudster to fly across the country, Davis was to lose interest in the company and resigned from it, taking the plane with him.

While the cross-country attempt was being developed, on 14 April 1921, the Davis-Douglas Company won its first order from the United States Government, Navy Contract No. 53305, which was worth US$119,550, for three experimental aircraft that were to be based on the original Cloudster design (the DT bomber). With Davis leaving the company, Douglas needed to establish his own company and looked to Bill Henry for assistance. Henry arranged for the publisher of the *Los Angeles Times*, Harry Chandler, to guarantee a US$15,000 loan to Douglas. This supplemented a small payment from the Navy for contract work on the DT bomber. In addition, Douglas also received a loan of US$5,000 from his father, who later became a member of the company's board. In July 1921, the company was incorporated as the Douglas Company, and as Donald Douglas and his father controlled a large part of the shares of the company, they pursued the strategy of reinvesting all but a small amount of the profits back into the company. This meant that by the fall of 1928, the net worth of the company had increased to US$2.5 million.

During the 1920s, Douglas built torpedo bombers for the United States Navy and observation planes for the Army (building more than 430 aircraft in the 1920s; Filucci, 2018 p. 93). It was, however, in the late 1920s' commercial aircraft revolution that Douglas really began to make his mark. This success meant that in November 1928, the company was reorganised and became the Douglas Aircraft Company Inc., a publicly traded company. A part of the money raised at this time went into construction a new aircraft factory at Santa Monica in California.

The Douglas Company became one of the leaders of the commercial aircraft industry in the United States, taking part in a decades-long struggle for dominance with its arch-rival the Boeing Company. To begin with, Douglas gained the upper hand, particularly with his revolutionary DC-1 and the successor aircraft that followed. The successful Douglas DC-3 airliner and its equally popular World War II military transport version, the C-47, were especially important aircraft in this regard. By the time World War II broke out Douglas' aircraft made up around 80 per cent of all commercial aircraft in service in the United States. Douglas' DC-3 proved to be the first airliner to make a profit by carrying just passengers, without the support of government subsidies or mail contracts or other forms. With this success, the company focused on the development of commercial aircraft. The Douglas DC-3 itself became one of the most important aircraft ever built and had the effect of revolutionising commercial aircraft travel. In doing so, the aircraft industry in the 1930s in the United States grew from a distant 41st place in employment size amongst American industries to first place in less than five years. Douglass Aircraft also grew at an extraordinary rate, from 68 employees in 1922 to the fourth-largest company in the United States.

After World War II was over Douglas continued in the industry, attempting to develop a range of new aircraft. His company, however, started to lag behind that of other companies in the jet age and was ultimately overtaken and surpassed by Boeing. Douglas himself stayed on after the war, retiring in 1957. He was replaced by his son, Donald Douglas Jr., as the company's president but Douglas Sr. retained his position as chairman of the board. Unfortunately, the company in the mid to late 1960s struggled to expand production to meet the demand for the DC-8 and DC-9 airliners and the A-4 Skyhawk military attack aircraft. Cash flow problems and DC-10 development costs, combined with shortages due to the Vietnam War, led Douglas finally to agree in April 1967 to a merger with the McDonnell Aircraft Corporation to form McDonnell Douglas. Douglas Sr. was made the honorary chairman of the McDonnell Douglas board, which he retained until his death at the age of 88 in 1981.

Douglas was unusual in that he actually saw the Wright brothers fly before their trip to Europe, an event which, like other Wright flights seen by aviation entrepreneurs, inspired him. His inspiration was fed by further demonstrations by Curtiss and his early model aircraft developments. He was an early adopter of the emerging university aeronautical education, which strengthened the technical skills learnt earlier at the military academy. After that, he chased opportunities in several companies to develop his knowledge until founding his own company in 1921, with the help of connections through his family. It could be argued that the inability of Douglas to keep up with technology developments and Boeing's aggressive market positioning ultimately led to the company falling out of a top manufacturing position. Quality, cash flow, development costs and an inability to meet demand continued to plague the company in the 1960s, suggesting that a model for the smooth running of the company had not been found. This difference was stark compared with the smooth operations of Boeing.

10.4 James Smith McDonnell

The Douglas Company eventually merged with another of its rivals, the company formed by James McDonnell. James Smith McDonnell (1899–1980) was an American aviator, engineer and businessman who was born in Denver, Colorado. McDonnell's father was a successful cotton merchant and store owner. James McDonnell was the fourth child in his family and he was raised in Little Rock, Arkansas, where he attended school and graduated from Little Rock High School in 1917. After that he then went to Princeton University in 1921, majoring in physics and taking an interest in aviation. Like Douglas, he went on to graduate from the unique Masters of Science in Aeronautical Engineering at MIT in 1925 (Pisney, 1981). At the same time, he broadened his knowledge of the aviation industry by joining the United States Army Air Corp Reserve where he received light plane training. After graduating from MIT, he was employed by Tom Towle at the Stout Metal Airplane Division of the Ford Motor Company. In 1927, he moved to the Hamilton Metalplane Company to develop metal monoplanes before moving onto the Huff-Daland Airplane Company. Despite being able to pursue his interest in the aviation industry, McDonnell seems to have been frustrated by working for others and hampered by the Great Depression, which engulfed the industry just as McDonnell was getting started.

McDonnell began to strike out for himself in April 1927, when he heard about a competition that was being offered by the Guggenheim Foundation for the successful design of a safe, light, training aircraft. To this end, the following year, McDonnell and two colleagues, Constantine Zakhartchenko and James Cowling, decided to leave their jobs with Hamilton, which at the time was being absorbed by UTAC, and organise a new company that became known as J.S. McDonnell Jr. & Associates. The three built a light aircraft known as the Doodlebug, but it wasn't able to win the Guggenheim competition. Around this time, the Great Depression began to strike the economy and McDonnell dissolved his firm in response and went to work firstly for the Great Lakes Aircraft Company in 1931 and then the Glenn L. Martin Company, where he became chief engineer for landplanes.

McDonnell seems to have similar plans to that of Shute Norway, in that the Doodlebug was a small, recreational aircraft. Recreation flying was a growing pastime in the 1920s but was largely snuffed out by the Depression. Unlike Shute Norway, he decided to see out the Depression years working for a more established surviving company, Glenn L. Martin, before re-creating his own company when the industry began to recover.

McDonnell eventually resigned from the Glenn L. Martin Company in 1938 and founded a new company, the McDonnell Aircraft Corporation, in the following year. To do so he raised US$165,000, principally from funds from his family and friends (Pisney, 1981, p. 5). By this time, demand for military aircraft was picking up, both domestically and abroad, and McDonnell decided to concentrate on this rather than his previous interest in recreational and commercial aircraft. After World War II McDonnell continued to develop military aircraft and became the country's most important producer of military aircraft. Headquartered in St.

Louis, the company quickly grew into a principal supplier of fighter aircraft to the United States Air Force and Navy, including the F-4 Phantom II fighter and the Mercury and Gemini space capsules.

The company's focus and reliance on military aircraft, however, led to some uneasiness amongst its management after World War II. In 1963, James McDonnell tried to gain control of the more commercial-aircraft-oriented Douglas Company, but at first was unsuccessful. Finally, in 1966, when Douglas was experiencing a serious financial crisis and needed cash desperately, McDonnell offered to buy Douglas stock, which allowed Douglas to obtain some much-needed cash. This measure also meant that the merger between McDonnell and the Douglas Aircraft Company took place to create McDonnell Douglas. Later that year Douglas Aircraft Company's space and missiles division became part of a new subsidiary called McDonnell Douglas Astronautics, located in Huntington Beach, California, producing the Delta series of launch vehicles. The new combined company also later developed the successful F-15 Eagle and F/A-18 Hornet fighters.

As well as running his company, McDonnell served as chairman of the United Nations Association of the United States, and in 1958, his company became the first organisation in the world to celebrate United Nations Day as a paid holiday. Later in 1980, McDonnell was awarded the NAS Award in Aeronautical Engineering from the National Academy of Sciences. He was succeeded as chair of McDonnell Douglas by his nephew, Sanford N. McDonnell, in 1980 and he died of a stroke on 22 August 1980, being buried in Bellefontaine Cemetery in St. Louis.

Like Boeing and Douglas, McDonnell came from a family of means, allowing him to pursue a university education and to later raise funds for the start-up of his companies. Unlike Boeing and Douglas, McDonnell's initial choice to pursue recreational and commercial aircraft was unsuccessful. The Great Depression wiped out the recreation market and others controlled the commercial market. Working for other companies allowed him to survive while giving him exposure to how other companies were operating and managing the downturn. When the threat of war re-emerged, McDonnell was well positioned to use his contacts and knowledge to build a company on the back of military contracts.

10.5 Conclusion

In the 1920s and 1930s, a number of aircraft manufactures turned to producing commercial aircraft instead of military ones (see Figure 10.2). In the case of the British ones, people like de Havilland and Shute Norway had some success with this strategy but suffered from the limitations inherent in British markets. American producers at this time found that with technological advances, it was possible to create production that sustained expansion and profitability.

The greater size of the American commercial domestic market meant that American manufacturers were able to market their products in the 1930s as passenger travel developed. Entrepreneurs like Loughhead, Douglas and Boeing were able to take advantage of the technological developments in the late 1920s

Figure 10.2 Passengers aboard an American airliner

and early 1930s to develop more advanced aircraft that could take advantage of this expansion. Late in the 1930s with the expansion of demand from the military, others like McDonnell were able to establish companies.

The level of education of this next generation of aviation entrepreneurs tended to be university based, rather technical or apprenticeship trained. They were, however, similar in that at an early age, they showed a fascination for flying and were prepared to get into, and stay in, the industry whatever the costs. They all showed a degree of restlessness while working for companies that they did not themselves control and usually attempted to create their own enterprises as soon as the circumstances made it possible.

As the industry became more complex, and economies of scale in production became more important, the larger market size became an important part of the development of the American industry. After World War II, competition for both military and commercial orders intensified. As the founders of the companies retired, a consolidation of the industry occurred. Even with this consolidation, some of the company founders stayed associated with the companies into the 1960s and 1970s.

References

Bauer, E. E. (1990). *Boeing in Peace and War*. Enumclaw, WA: TABA Publishing.

Collins, J. C. (2001). *Good to Great*. New York, NY: HarperCollins Publishers.

Collins, J. C. and Porras, J. I. (1995). Building a visionary company. *California Management Review*, 37(2), pp. 80–98.

Collins, J. C. and Porras, J. I. (1997). *Built to Last*. New York, NY: HarperCollins Publishers.

Cunningham, F. (2010). *Skymaster: The Story of Donald Douglas and the Douglas Aircraft Company*. Philadelphia, PA: Kessinger Publishing.

De Havilland, G. (1961). *Sky Fever*. London: Hamish Hamilton.

D'Intino, R., Boyles, T., Neck, C. P. and Hall, J. R. (2008). Visionary entrepreneurial leadership in the aircraft industry: The Boeing company legacy. *Journal of Management History*, 14(1), pp. 39–54.

Filucci, J. B. (2018). *Honest Vision: The Donald Douglas Story: Timeless Leadership Lessons from an Engineering Mind and Aviation Icon, Aviation Supplies and Academics.* Annapolis, HD: Naval Institute Press.

Francillon, R. J. (1988). *McDonnell Douglas Aircraft Since 1920*, Vol. 1. Naval Institute Press. Annapolis, MD: Naval Institute Press.

Mansfield, H. (1966). *Vision: The Story of Boeing: A Saga of the Sky and the New Horizons of Space*. New York, NY: Popular Library.

Morrison, W. H. (1991). *Donald W. Douglas: A Heart with Wings*. Ames, IO: Iowa State University Press.

Pisney, R. F. (1981). *James S. McDonnell and his Company: A Vision of Flight and Space*. St. Louis, MO: Missouri Historical Society.

Redding, R. and Yenne, B. (1983). *Boeing: Planemaker to the World*. Greenwich, CT: Cresent Books-a Bison Book.

Rodgers, E. (1996). *Flying High: The Story of Boeing and the Rise of the Jetliner Industry*. New York: The Atlantic Monthly Press.

Sarasvathy, S. D. (2009). *Effectuation: Elements of Entrepreneurial Expertise*. Cheltenham, UK: Edward Elgar Publishing.

Serling, R. (1992). *Legend and Legacy: The Story of Boeing and its People*. New York: St. Martin's Press.

Part III

Summary

11 A global industry

11.1 Introduction

On 30 December 2016, the United States Air Force issued a formal request for proposals to supply a trainer jet aircraft to replace its aging trainer fleet. One very interesting aspect of the various competitors that submitted bids to supply this very lucrative contract was the range of international partnerships. These included the Italian company Leonardo S.p.A. with the M-346-based T-100; Korea Aerospace Industries/Lockheed Martin with the T-50 Golden Eagle; the Boeing and the Swedish Aircraft Manufacturer (Saab) Group with a new, purpose-built Boeing T-X; and Sierra Nevada Corporation and Turkish Aerospace Industries with a new design. Northrop Grumman and the British company BAE Systems decided to withdraw from its proposed bid.

What's incredible about the process is the degree of international cooperation and pooling of resources that takes place to create the best possible product for the United States Air Force. Tracing the development of this international cooperation is a substantial story in itself, and one that dates back a long way in the industry. One remarkable aspect of the early aviation industry was the global interaction that took place between the aviation pioneers, not only in the United States and Western Europe but also other countries such as Russia, Italy and Japan. The global aviation industry is often one depicted as being dominated by government military purchases, government-owned and operated carriers and owned or subsidised national aviation defence industries. It is true that all these have in the past existed but, nonetheless, the global nature of the industry from the very early days is remarkable.

The purpose of this chapter, therefore, is to take a look at some of the global interactions of the aviation pioneers, and therefore to better understand the dynamics of the industry. In addition, some material will be presented on the present-day aviation industry.

Although it is not possible in the short space of a single chapter to provide a description of all the ways in which the aerospace industry is a global one, some aspects of it are described to provide some indication of it. The examples provided will tend to take in organisations that have survived, in one form or another, over the industry's long history.

11.2 A global industry

Determining the scale and scope of the aerospace industry today is a little tricky as many of the major players combine aerospace with a range of other industries. An aerospace manufacturer is generally regarded as being an individual or firm that is involved in the various aspects of designing, building, testing, selling and maintaining aircraft, aircraft parts, missiles, spacecraft or rockets. The industry is, therefore, a high-tech one. In addition to manufacturing aircraft there is also the manufacturing of parts for aircraft used in maintenance, both for civil and military aviation.

In 2017, the global aerospace industry was worth US$838 billion (see Table 11.1), with aircraft and engine original equipment manufacturers (OEMs) representing 28 per cent of the total (US$235 billion), civil and military maintenance, repair, overalls (MRO) and upgrades 27 per cent (US$226 billion), aircraft systems and component manufacturing 26 per cent (US$218 billion), satellites and space 7 per cent (US$59 billion), missiles and unmanned aerial vehicles (UAVs) 5 per cent (US$42 billion) and other activity, including defence electronics, flight simulators and public research accounting 7 per cent (US$59 billion) (AeroDynamic Advisory/Teal Group Corporation, 2018).

The nations with the largest sectors are the United States with US$408.4 billion (49 per cent), followed by France with US$69 billion (8.2 per cent), China with US$61.2 billion (7.3 per cent), the United Kingdom with US$48.8 billion (5.8 per cent), Germany with US$46.2 billion (5.5 per cent), Russia with US$27.1 billion (3.2 per cent), Canada with US$24 billion (2.9 per cent) and Japan with US$21 billion (2.5 per cent) (Table 11.1). Although much of the industry is concentrated in the United States, in a number of the other cases (e.g. France, the United Kingdom and Russia), aerospace exports are the countries' largest manufactured exports, as is the case with the United States. This means that the industry plays an important part in the country's economic development.

Table 11.1 The global aerospace industry, revenue in US$billion, 2017

	$b	*Per cent*
United States	408.4	49
France	69.0	8
China	61.2	7
United Kingdom	48.8	6
Germany	46.2	6
Russia	27.1	3
Canada	24.0	3
Japan	21.0	3
Spain	14.4	2
India	11.0	2
Other	106.9	1
Total	838.0	100

Source: AeroDynamic Advisory/Teal Group Corp (2018)

Table 11.2 provides a list of the world's 20-largest aerospace firms in 2017. As can be seen from the list, most are headquartered in the United States, although it is noteworthy that many of these American companies have substantial operations outside of the United States. In the United States itself, NASE and the Department of Defense are the two largest consumers of aerospace technology and products, along with the commercial airlines. The United States Bureau of Labor Statistics reported in 2016 that the aerospace industry employed 490,000 people, many of which were in Washington and California (Aerospace Industries Association, 2017).

In addition to the American companies, a number of companies based in the European Union are important, such as BAE Systems, Airbus, Dassault, Thales, Saab AB, Terma A/S and Leonardo. In Russia, there are also large aerospace firms such as the United Aircraft Corporation (encompassing Mikoyan, Sukhoi, Ilyushin, Tupolev, Yakovlev and Irkut, which includes Beriev) and Oboronprom. A number of these companies (both American and others) are the heirs of those established by the aviation pioneers mentioned in this book.

In recent decades direct government involvement in the industry has declined, as commercial aviation carriers have been privatised along with defence manufacturers. Despite these moves, there is still a strong relationship between government and the industry through the purchases of the former for aerospace equipment for the military. The heavy relationship between military purchases and the aviation industry has meant that direct government involvement has a long history in the industry.

Table 11.2 Twenty largest aerospace companies in the world by revenue, 2017

Company	*Revenue $b*	*Profit $b*	*Country*
Boeing	94.6	4.90	United States
Airbus	70.8	2.40	European Union
Lockheed Martin	47.2	5.55	United States
United Technologies	29.0	3.84	United States
GE Aviation	26.3	6.12	United States
Northrop Grumman	24.5	3.19	United States
Raytheon	24.1	3.24	United States
Safran	16.8	2.54	France
BAE Systems	13.4	–	United Kingdom
Leonardo (Finmeccanica)	12.8	1.05	Italy
Rolls-Royce Holdings	12.0	0.98	United Kingdom
Honeywell	11.7	0	United States
L-3 Technologies	10.5	1.0	United States
Textron	9.9	1.0	United States
Bombardier	8.8	−0.5	Canada
Mitsubishi Heavy Industries	8.7	−0.2	Japan
General Dynamics (Aerospace)	8.4	1.7	United States
Precision Castparts	7.0	0	United States
Spirit Aero Systems	6.8	0.7	United States
Embraer	6.2	0.2	Brazil

Source: Flight Global Archive

Airline businesses were small in the 1920s and 1930s, but in the post-World War II period grew to a substantial size. In the late 1950s, the wide-bodied large capacity aircraft that were developed had a great impact on the volume of public and product transported. In the case of the United States in the 1930s, open access to the skies in the interwar period allowed for the development of several airlines carrying passengers, mail and freight across the country. In the European case, by the late 1940s, the picture of European travel was of each country having one monopoly carrier with airspace tightly controlled by each nation state. The 1919 Paris Air Convention endorsed each country's sovereignty over its airspace.

This occurred because airspace was seen as having significant military value so that each country wanted to ensure it held exclusive property rights to it. Airlines were also seen as a means of communications with potential military and social significance. At a minimum that led to government subsidies and often government ownership of airlines, and sometimes aerospace manufacturing. The new industry saw many fledgling companies and because of the limited opportunities to expand at will on international routes, many stuck to domestic markets, which became overcrowded, except in America, where longer distances made air travel popular. This meant that outside America, the firms became unprofitable, and in the 1920s, were generally bankrupt. To maintain service, governments began to subsidise companies for political and military reasons (even in the American case this was done through mail contracts and subsidised airport facilities). Mergers were common and generally led to the creation of a single company as the nation's chosen instrument with a state subsidy. The subsidised Air France, for instance, was established in 1933 with 25 per cent of shares state owned. In the United Kingdom, foreign companies were banned, and in 1935, British Airways was established with government subsidy from the merger of several unprofitable companies. It mainly operated within Europe, while Imperial Airways, which was set up in 1924, provided services to India, South Africa and Australia.

After World War II, these trends became institutionalised in the form of bilateral agreements between countries, and in the French and British cases, they were used to foster local aircraft manufacture. This process reduced the potential for tourist traffic to grow outside of the United States before the 1960s, even though the improvements in airports, navigational aids, and bigger aircraft with pressurised cabins, bigger payloads and longer travel distances encouraged growth. The period from the late 1950s to the mid-1970s, therefore, saw major technological changes, even while the institutional arrangements stayed intact.

Eventually, during the 1960s and 1970s, the growth of charter inclusive tours helped to break up the system. Government monopolies were abolished and in the case of the United States markets eventually deregulated (1978). Downward pressure on costs and prices as aircraft became bigger and more sophisticated pushed the industry forward (Lyth, 1998; Dierikx and Lyth, 1994; Brooks, 1967). Despite all of this heavy involvement by governments, international links were important in the early years of aviation.

11.3 Early global efforts: exhibitions and contests

To a large degree, the global interaction between the pioneers predates the first flight by the Wrights in 1903. In the lead up to this event, a number of developments occurred, including the international development of combustion engines, but also the developments of glider flights by people such as Lawrence Hargraves in Australia and Otto Lilienthal in Germany. The aviation pioneers in the United States, France and the United Kingdom took notice of these developments and used the knowledge from them to develop their own plans.

Secrecy and scepticism about their work meant that Wrights initially struggled to sell their planes to military interests in the United States and Europe. Public flight demonstrations were needed, resulting in Wilbur demonstrating in Europe, while Orville flew near Washington, DC.

Wilbur's flights at Le Mans, France, changed public perception for ever, and the general public, particularly in Europe, fell in love with the Wright Brothers and aeroplanes.

The Wrights' success encouraged others to take to the air. Air shows and prizes grew in number and size, attracting international contestants. In August 1909, for instance, Glenn H. Curtiss took part in the Grande Semaine d'Aviation aviation meeting at Reims in France, organised by the Aéro-Club de France. The Wrights decided not to compete in person, as they were selling their machines to customers in Germany at the time, although two Wright aircraft were at the meet, but did not win any events. Instead, flying his No. 2 biplane, Curtiss won the overall speed event, the Gordon Bennett Cup, completing the 20-kilometre course in just under 16 minutes, beating the runner-up Louis Blériot.

The Grande Semaine d'Aviation de la Champagne was the first international public flying event and was seen both at the time and by later historians as marking the coming of age of heavier-than-air aviation. Most of the prominent aviators of the day decided to take part and the 500,000 people visited the meet, including Armand Fallières, the president of France, and the British chancellor of the Exchequer and later prime minister David Lloyd George.

This event stimulated further prizes designed to push the technical limits of flight. In England, the *Daily Mail* offered a prize in October 1908 of £500 for a flight made before the end of the year across the English Channel. When 1908 passed with no serious attempt being made to win the prize it was doubled to £1,000 and the offer extended to the end of 1909. Louis Blériot took up the challenge to win this prize and on 25 July 1909 made the crossing. Other exhibitions, racers, challenges and prizes followed, which increased the level of international involvement.

Not all of these prizes and competitions were European based. In the United States, William Randolph Hearst was particularly interested in the newly emerging technologies relating to aviation and had his first experience of flight in January 1909 in Los Angeles. He sponsored the Hearst Transcontinental Prize and a number of European flyers tried, without success, to win it. One of these was the

English flyer Thomas Sopwith. On 18 December 1910, Sopwith won a £4,000 prize for the longest flight from England to the continent in a British-built aeroplane, flying 272 km in 3 hours 40 minutes. He used the winnings to set up the Sopwith School of Flying at Brooklands, which in turn begat the Sopwith Aviation Company. In 1911, he travelled to the United States seeking more prize money with mixed success. Much later in life in the mid-1930s, he returned to the United States on two occasions to seek the America's Cup, the symbol of American yachting excellence.

The general excitement over the beginning of flying in a number of countries meant that there was a great deal of observation of the developments taking place. Prizes, exhibition, flying and races helped generate cross-country interest that was to extend into the interwar years, resulting in first flights across the Atlantic and the Pacific, from England to Australia and the iconic Schneider Trophy.

11.4 Manufacturing under licence

International exhibitions and contests brought to the attention of the public and aviators the different levels of development of the industry across countries. In the earliest years, a number of countries struggled to keep up with the aviation leaders in the United States and France. This meant that licencing production offshore became common. Pre-World War I, sales were hard to come by, and so aviation pioneers were attracted to this form of sale.

In Europe in the pre-World War I period, Germany was numerically the leader of this the group, with some 25 aircraft production companies established before World War I. Many of these manufacturers, however, began by obtaining licences to produce foreign designs rather than design their own aircraft. In 1910, E. Rumpler was the first of these, and he was licenced to produce the Austrian *Taube* (Dove) at his factory. Other German companies, such as Albatros, began production in 1910 with French designed aircraft, later adding the Taube-type monoplanes to its production line. Aviatik, founded in 1910, also began by manufacturing two French aircraft. In addition, the German Government supported the creation of the Deutsche Flugzeug Werke, which produced the Maurice Farman biplane in 1910 and later produced a monoplane that won the 1913 Prince Henry trophy. This company gradually developed its own aircraft, but to begin with made foreign-designed aircraft. Another who followed this course was August Euler, who established the Euler works and began manufacturing Voisin biplanes under licence. Finally, Gustav Otto began building aircraft under licence with Henry Farman and then later developed his own Otto tractor biplane.

In the United Kingdom, companies manufactured their first planes from French designs. Bristol, for instance, designed its first aircraft as a licenced and improved version of an aircraft manufactured in France by the Société Zodiac. This was a biplane designed by Gabriel Voisin. Along with French designs, the Wrights' Flyer was manufactured under licence by various companies in the United Kingdom and other countries in Europe.

The surge in developments in France and the United Kingdom, both just before the outbreak of war and after, meant that at this stage the American industry was left behind. When it entered the war the United States Government, therefore, decided that it would focus its efforts on the manufacture of a single European aircraft. In May 1917, the War Department sent the Bolling Commission, headed by Raynal C. Bolling of United States Steel, to visit Europe in order to inspect the aviation industry and to decide which aircraft to manufacture. The aircraft that was chosen was the British DH-4, designed by Geoffrey de Havilland. In making this choice they were influenced by the British Government granting free use of its licence for the aircraft. This was in contrast to the French aircraft, which required payment of a royalty, even though in some cases they were superior aircraft. Not everyone in the United States favoured this course of action. The industry pioneer Groever Loening, for instance, argued that there were promising American aircraft designs such as the Martin MB-1 bomber, the Vought trainer, the Loening M-8 and the Thomas-Morse MB-3. These, however, were ignored in favour of the foreign designs. Critics at the time also accused the companies that had won the contract of inflating costs in order to increase profits. Amongst the largest of the DH-4 construction contracts were the companies Dayton-Wright, Glenn L. Curtiss, Fisher Body and Standard Aero in New York. Despite the efforts made by these companies to make the DH-4, the number actually shipped to Europe was small. In the end, most United States airmen in Europe flew French-made aircraft.

At the same time, British personnel were also employed in American factories. One example was the case of Curtiss, who met two individuals who would prove to be important to his enterprises. The first was B. Douglas Thomas, who joined the Curtiss Aeroplane and Motor Corporation and worked on the design of the Curtiss J and N models. The second was a Royal Navy pilot, John Porte, who would later be one of the pilots who crossed the Atlantic in a Curtiss seaplane.

As well as airframes, World War I also meant that there was a steady need for engines. In the United States, to supply them, the investors in the Wright Company bought an interest in the Simplex Automobile Company. After a trip to France to study engines, they decided to build the French designed Hispano-Suiza engine in the United States under licence from the French Government. Wright received a contract to build 450 of them.

This was not the last example of the American manufacture of European aero engines. Most famously the Rolls-Royce Merlin was manufactured by Packard in the United States during World War II and used to power the world-record-beating Mustang fighter. Less well known is that Curtiss-Wright closed its Airplane Division in March 1951 and focused on engine and propeller manufacturing. Curtiss propellers went on several major airliners, and the R-3350 engine evolved into an efficient power plant, which was used until jets became dominant. A Wright engine powered the DC-7. The company's first foray into the jet age, however, was with the production of the British Sapphire jet engine beginning in 1952, which was manufactured under licence as the J65. It powered a number of American-built military aircraft.

From time to time other European aircraft have also been built in the United States besides the DH-4. One example of this is when Fairchild, in 1956, purchased the licence to produce the Fokker F-27 Friendship passenger aircraft, making approximately 200 of six versions as well as its own stretched FH-227 version, which first flew in 1966. The F-27 became the first American-built jet airliner in service and, along with the FH-227, became widely used as feeder planes for commercial carriers both in the United States and abroad.

Perhaps the most unique European aircraft produced in the United States was the Hawker Siddeley Harrier 'jump jet', which could take off and land vertically, but which flew like a conventional plane. McDonnell Douglas, British Aerospace and Rolls-Royce developed the AV-8B Harrier II, which in turn was based on the 1957 British-designed Hawker Siddeley Kestrel. Production in the United States began in 1981, and more than 340 Harriers were produced, mainly for the Marines. In addition, the Martin B-57 Canberra was an American-built, twin-jet bomber and reconnaissance aircraft that entered service with the United States Air Force in 1953. The B-57 was a licence-built version of the British English Electric Canberra, built by the Glenn L. Martin Company. Initial Martin-built models were virtually identical to their British-built counterparts, although Martin later modified the design to incorporate larger quantities of United States sourced components. The American-built B-57 Canberra was sold to export customers and combat use was seen by the Pakistani Air Force during the Indo-Pakistani War of 1965 and the Indo-Pakistani War of 1971.

In more recent years the McDonnell Douglas (now Boeing) T-45 Goshawk was produced as a highly modified version of the British BAE Systems Hawk land-based training jet aircraft. The T-45 is used by the United States Navy as an aircraft carrier-capable trainer. Production of the Goshawk was divided between the two primary partner companies, with BAE Systems responsible for manufacturing the fuselage aft of the cockpit, along with the air inlets and the vertical stabiliser of the T-45, along with the wings. McDonnell Douglas and then Boeing undertook the manufacture of the remaining elements of the Goshawk, as well as conducting assembly of the type at a production line at St. Louis, Missouri. The Hawk is one of a few aircraft designed back in the 1960s, which is still being produced today. The Hawk and another example of aircraft from the 1960s still in production are provided in Figures 11.1 and 11.2.

Although these examples of European designs being produced in the United States are important, it has been more common for American designs to be built elsewhere. The 1937 Lockheed 14 Super Electra, for instance, was designed to compete with the Douglas DC-2 and DC-3 but failed as a commercial aircraft in the United States because it had too small a capacity. Most were sold abroad, with more than 100 being licence-built in Japan for use by the Imperial Japanese Army. The 14 Super Electra formed the basis for the Lockheed Hudson, which was used by the United Kingdom's Royal Air Force in World War II. Its rival, the DC-2, was also constructed under licence by a foreign manufacturer, by the Norwegian Government and later by Peru's Navy. In the case of the DC3, Douglas built a total of 10,655 of them and about another 2,500 planes were built under licence in the Soviet Union and Japan.

Figure 11.1 WinAir de Havilland Canada DHC-6–300 Twin Otter at St. Barthelemy Airport

Figure 11.2 T-45A Goshawks training aircraft cruise 4 June 2008

Along with commercial aircraft, it was common to see American military aircraft built outside of the United States. In Germany, for instance, Messersmitt's company produced an advanced American fighter, the Lockheed F-104, in the 1960s. Heinkel also found work by assembling aircraft of foreign design under licence, including the F-104.

In Japan, Mitsubishi, a family-owned industrial combine that owned banks, along with other ventures, was founded in 1873 and received strong government support from the outset. It started with trading and merchant shipping, then it expanded into mining, shipbuilding, real estate, iron and steel, insurance, oil refining and chemicals. The group entered aviation during World War I, and at first manufactured French aircraft engines under licence. A design group headed by the United Kingdom's Herbert Smith, who had been a chief engineer at the Sopwith Aviation Company, built new aircraft that became standard equipment with the Japanese Navy. Mitsubishi also learnt lessons from Germany, first by working with the aircraft designer Alexander Baumann and then through a collaboration with the German company Junkers. After World War II, Mitsubishi went on to assemble American jet fighters, building them under licence, and later supplied sections of airframes for the Boeing 767 and 777 and McDonnell Douglas MD-11 airliners.

In Sweden, the aerospace industry has often built modified versions of United States or British designs under licence. Flygmotor, a branch of Volvo, built some 1,200 engines under licence to overseas manufacturers during World War II. At the same time, Saab constructed nearly a thousand aircraft, as bombers, fighters and reconnaissance craft, mainly using the Flygmotor engines. Saab's first post-war fighter was the jet-powered Tunnan and flew with a British de Havilland Ghost Engine that Flygmotor again built under licence. Around 660 Tunnans were delivered to the Swedish Air Force between 1951 and 1956. Next came the Lansen, which used another British engine, the Rolls-Royce Avon, again manufactured under licence. Sweden entered the supersonic era with its Draken, which used the licence-built Avon.

In the wake of World War II, the French industry relied quite a bit on foreign designs. Aérospatiale, which began in 1950, produced the Languedoc airliner and two fighters, the Vampire and Mistral. The Vampire was British, being built under licence; the Mistral used a jet engine, the Nene, from the United Kingdom's firm of Rolls-Royce. Avons were also used in an early French jet airliner, the Caravelle.

The list of aircraft built under licence in other counties given here is by no means exhaustive, with there being many other examples of this occurring over the years. In some cases, national industries have made use of overseas production types for nearly all of the aircraft built in that country, a good example being that of Australia, which has seen the construction of a succession of British, American and French types.

The manufacture of aircraft and parts is today still undertaken under licence along with the creation of numerous subsidiaries of the major companies in countries around the globe. The spread of manufacturing, therefore, is more varied than the figures in the table would indicate, and this spread has been a feature of the industry since its foundation

11.5 Engines

One constant since 1903 has been the global nature of aircraft engine production. At the start of the twentieth century, aero engines were fairly simple, low-powered machines that were designed and built for each specific aircraft. Soon engines began to be built in quantity, sometimes by several manufacturers in different countries. In the United States, particularly during World War I, motorcar manufacturers dominated the manufacture of aero engine until firms that specialised in aero engines were established in the 1920s.

To begin with, aero engines were stationary, either radial in style or in-line. The French Antoinette series was the most commonly used in early aircraft. These were followed by the popular rotary engine, the best known of these being the Gnome and Le Rhône, which were used on the majority of aircraft until the in-line Liberty engine, designed for mass production, started dominating the aero engine market. From that point on, increasingly more sophisticated and powerful stationary in-line engines were developed until the arrival of the jet engine a couple of decades later.

The Antoinette engine was designed and built in France by Léon Levavasseur and named after the designer's daughter. It was Europe's most widely used engine up until 1910. The first Antoinette engine dated from about 1901 and was used to power a speedboat. The Gnome 50-h.p. rotary engine, which followed, was to revolutionise aviation. Although F.O. Farwell developed the first successful air-cooled rotary engine in the United States in 1896, it was the French-built Gnome rotary that was first used extensively in aircraft during the first years of World War I. It was used I primarily in the Nieuport 28, which the United States Air Service purchased from France, and also by the thousands in other Allied aircraft. This engine, produced under the names of Bentley Rotary B.R.1 in the United Kingdom, Thulin in Sweden and Oberursel UR.I in Germany, dominated the industry until 1916.

Around 1911, other engine manufacturers began building rotary engines. LeRhône and Clerget engines, both built in France, were used in many Allied fighter planes in World War I. LeRhône engines were quite reliable, and by the end of the war, they were being manufactured in the United Kingdom, Italy and the United States, in addition to France. The Germans also produced the 110-h.p. Oberursel, which was practically an exact copy of the 110-h.p. LeRhône.

A Swiss engineer who worked for a Spanish car company, Marc Birkigt, designed a number of Hispano-Suiza products, which became popular. These included a large number of water-cooled V-8 and V-12 engines that were made in France, the United Kingdom and the United States during the war. They were especially compact for the amount of power they produced.

By far the most important of all Allied engines, and the United States' most significant contribution to the war effort, was the Liberty. On 4 June 1917, the Aircraft Production Board authorised its final design and manufacture. Assembly of the first eight-cylinder version was completed in the remarkably short span of less than six weeks, and the Liberty became the standard wartime aircraft engine, produced by Packard, Lincoln, Ford, General Motors (Cadillac and Buick), Nordyke and Marmon. More than 13,000 engines came off the assembly line before the Armistice and later it was exported to the United Kingdom in World War II and used to drive tanks.

At the same time that Liberty engines were being sent to the United Kingdom in World War II to drive tanks, Rolls-Royce Merlin engines were being produced in the United States. Rolls-Royce began producing aero engines during World War I after the Royal Aircraft Factory asked Rolls-Royce to develop a new 200-h.p. air-cooled engine. In a much later era, Rolls-Royce became once again involved in the American industry when in 1995 it acquired the Allison Engine Company, a manufacturer of gas turbines and components for aviation, industrial and marine engines. The two companies had a technical association dating back to World War II and Rolls-Royce had previously tried to buy the company when General Motors sold it in 1993, but General Motors opted for a management buyout instead for US$370 million. In 2001, Rolls-Royce and its LiftSystem was amongst the group that won the JSF contract for the F-35.

The production is just one example of the complex components that go into the production of modern aircraft. Throughout the history of the industry, a substantial international trade has been undertaken in engines and other aircraft components, which has added to the global nature of the industry.

11.6 Conclusion

Although aviation is generally seen as being a global industry because of the importance of international commercial air travel, the global nature of the industry became apparent before commercial air travel became possible. Indeed, the 1908 contests between the likes of Blériot and Curtiss seem as much as anything to mark the beginning of a tremendous international interest in the industry as well as the beginning of an international network of flyers and manufacturers which had continued down to the present day. A key element of the success of the aviation industry has been its ability to share and build on others' designs and technology quickly.

All of the early pioneers of flight had an interest in international events and this interest continued as they developed their manufacturing concerns. Although military interests and government involvement might have had the effect of dividing the industry along national lines, the global interlinkages continued into the present day. These links exist today in a variety of forms, many which have a long history stretching back to the earliest days during World War I.

References

AeroDynamic Advisory/Teal Group Corporation. (2018). *The Global Aerospace Industry: Size & Country Rankings*, 16 July.

Aerospace Industries Association. (2017). *Facts & Figures: U.S Aerospace & Defense*, Arlington, VA: AIA.

Brooks, P. W. (1967). The development of air transport. *Journal of Transport History*, 1(2), pp. 164–183.

Flight Global Archive. Available at: www.flightglobal.com/pdfarchive/view/1909/1909%20-%200255.html

Lyth, P. J. (1998). *Flying the Flag: European Commercial Air Transport Since 1945*. Basingstoke, UK: Palgrave Macmillan.

Lyth, P. J. and Dierikx, M. L. (1994). From privilege to popularity: The growth of leisure air travel. *Journal of Transport History*, 1592, pp. 97–116.

12 Starters and runners

12.1 Introduction

After looking at so many of the early aviation entrepreneurs it is possible now to gain a better appreciation of a number of their personal characteristics, which appear in many of them, as well of the opportunities and problems that they faced. Although it is not necessarily true that they all possessed identical characteristics, it is true that they often behaved in similar ways, such that we can compare them. It also should be borne in mind that not all of them entered the industry in the same manner, and at the same time. Those that entered at the start of the aviation industry and during the interwar years tended to be less successful than those who started in the lead up to the world wars. This is not surprising. In the very early stages, the military could not see the potential of aviation, but once it did, it was keen to get its hands on it. Despite these differences in context, certain similarities in entrepreneurial behaviour can be seen across the group chosen. Though not identical, nonetheless there does still appear to be some similar behaviours on the part of a number of them. From these behaviours it is possible to build up knowledge of some aspects common to entrepreneurs in general.

In addition to personal behaviours, the context of the industry's development is also crucial. Entrepreneurs seem to thrive most in a business climate where technical and market change is rapid, and this was true of the aviation industry. Engagement by the military opened up the industry, creating much-needed opportunities and money for those that could meet military needs, making it easy for these start-ups to survive and thrive. However, significant technological changes occurred between World War I and World War II, which created an industry where economies of scale were important, such that a limited number of large companies eventually dominated the industry and effectively made it difficult for individual entrepreneurs to prosper. Context, therefore, is important. Another source of context is the nature of demand for aircraft. Unlike, say, the motorcar industry, the aeronautical industry had a limited number of sources of demand. Aircraft use never became the popular recreational pastime that driving motorcars did, and so much of the demand for aircraft remained limited to the military and to a small number of commercial airlines. This concentrated structure of demand eventually encouraged a concentration in company structures. The roller coaster pattern of supply and demand between 1910 and 1959 also led to several expansion and contraction periods that stimulated innovation and tested the durability of entrepreneurs. After each world war, military money was effectively switched off, meaning that for entrepreneurs to survive this desert period they had to move quickly to adapt or shift the focus of their businesses.

It is possible to note the nature of the industry itself. Aviation in its earliest days attracted particularly adventurous individuals and it is possible to claim that the inherent danger of early flight made the industry different in nature to the high-tech industries of other periods, including today. The view in this work, however, has been that the aviation industry was not fundamentally different from other industries, although perhaps early aviation entrepreneurs did need a heightened degree of propensity for risk-taking, particularly during the initial formation stage of the industry. It is really a matter of degree of risk-taking that is different between the aviation pioneers of the past and the high-tech innovators of today. The latter group risk only their finances, and perhaps their reputations; the former certainly risked that as well, but also their lives.

Finally, how entrepreneurs learn and convert this experience to knowledge is particularly important to consider in high-tech dynamic, emerging industries. During start-up and highly dynamic phases of the industry, established educational institutions have not yet formalised the emerging knowledge into a structured educational offering. Communities of practice, therefore, can and often do play a critical role in bridging and filling gaps in knowledge.

The purpose of this chapter, therefore, is to take a broad, overall look at the aviation entrepreneurs presented in this book to see what personal behaviours were common to them. In addition, some of the company structures that they created are looked at, and in the final section, some conclusions are made.

12.2 Personal behaviours

Entrepreneurs' behaviour plays a key role in the evaluation, discovery and exploitation of opportunities fueling the emergence and growth of business enterprises (Mueller et al., 2012). In the earliest years of heavier-than-air flight, most of the aviation manufacturing companies that were established in the lead up to World War I were created by the pioneers of flight itself. Men like the Wright brothers and Glenn H. Curtiss in the United States, A.V. Roe and Thomas Sopwith in the United Kingdom and Louis Blériot in France are all examples of this. There were a few less common exceptions to this trend, such as the companies established with the capital and expertise of other larger organisations – for example, Vickers, Armstrong Whitworth, Bristol and the Aircraft Manufacturing Company in the United Kingdom and Boeing in the United States (with his timber interests).

The reasons for this are varied, but perhaps the most important was simply that in its earliest days, the aviation pioneers were the only ones who possessed the expertise needed to construct aircraft in an industry, bearing in mind the degree of technological change and the lack of knowledge and skill in the field. This made it difficult for established firms in engineering, for instance, to enter this field as they found it difficult to attract those with knowledge in this emerging field to work for them. This is an aspect that seems relevant to us today when looking at the emergence of new industries. In newly emerging industries, it is often difficult for established companies in related fields to enter as they have trouble understanding the challenges of the emerging industry. They also have difficulties in attracting the skills and knowledge of those able to address these challenges. This adds to the high-risk profile of engaging in or with start-up ventures in emerging markets.

The pioneer aviators/entrepreneurs all, almost without exception, had an enthusiasm for flying itself and had mechanical and design skills, which helped them to make a mark for themselves in this emerging field. Each admired a previous aviator/inventor – the Wrights were great admirers the German inventor Otto Lilienthal, whereas later entrepreneurs took the Wrights themselves as their inspiration. Each aviation entrepreneur built on the knowledge of these pioneers, adding to and deepening understanding of aviation flight and its practical application. Their learning was therefore driven by a focus on the promise of the future, an 'act now, experiment quickly' approach which utilised whatever knowledge they had learnt from the past or were currently learning.

Another common attribute common to these aviation entrepreneurs was a tendency towards a lack of patience for pursuing a conventional career in industries such as engineering. Most of them came out of related industries such as the manufacture of bicycles, motorcars, motorbikes or electrical engineering, but nearly all of them were young, adventurous types attracted to the rigour, speed and excitement of flight. This adventurousness contributed to a restlessness in and reluctance to fully engage in a conventional engineering career. In an age when most people stayed in jobs for long periods of time, most of our entrepreneurs drifted from one job to another before settling down in aviation. In the American case, self-employed types like the Wrights and Curtiss are common, although ones like Sopwith also existed in the United Kingdom. A number also moved in and out of jobs in related industries to fund their company start-ups or to weather the downturn periods when orders and money were in short supply. Again, this seems relevant to us today in looking at new high-tech industries.

All of the aviation pioneers appear to have possessed adventurous (even dangerous) hobbies, most importantly flying, but in their youth, bicycle, motorbike or car riding. Often later in life their attentions turned to pursuits such as yachting and wildlife photography. Many, like A.V. Roe and Curtiss, were winners of bike riding races and record holders in speed races, tweaking existing machines to enhance performance. The Wrights, Roe and Curtiss were also compulsive inventors from a very early age, showing a proclivity for the invention of new products as well as the ability to acquire patents in their childhood.

Being at the cutting edge of emerging technology, these aviation entrepreneur interests lay in pushing the limits of learning, resulting in higher order learning that allowed them to create new frames of reference (Huber, 1991), that led them beyond basic skill development to the new insights needed to, in the early stages, outperform others in their hobby areas, and later to fly (Fiol and Lyles, 1985). In a number of cases, personal and business crises drove, rather than diminished, their pursuit of higher order learning, resulting in the further development of technology solutions to aviation flight problems.

In terms of their social and economic backgrounds these entrepreneurs often came from middle-class backgrounds, but at an early age showed a proclivity for technical pursuits, rather than the interests associated with an academic education. In the British and French cases, most attended technical schools or colleges before going into heavy or light industry. A number of the British pioneers came out of the technical college system and completed engineering apprenticeships, which provided them with the skills and knowledge that helped them to work in aviation. Later, entrepreneurs such as Blackburn and Shute Norway in the United Kingdom and Boeing, McDonnell and Douglas in the United States had university engineering degrees; however, the earliest entrepreneurs tended not to. This is unsurprising as in the early stages of the industry, aviation knowledge was still under construction and rapidly changing, making it impossible for universities to acquire and package this knowledge into courses. Eventually, American entrants attended institutions like MIT, which was especially important in preparing people for a career in aviation. University educated men like Blackburn and Shute Norway were eventually attracted to the industry, mainly after undertaking engineering degrees. In terms of their education, therefore, the early innovator entrepreneurs tended not to have conventional university style backgrounds, being more interested in accruing their experience and knowledge from related industries such as bicycle/motorcycle manufacturing or the motor trades.

In most cases, these entrepreneurs had the financial support of their families, who often supported their choice of technical education and provided financial assistance to get their start-up ventures going. For example, family finance supported the start-up ventures of the Roes and McDonnell. They were also occasionally able to attract investment funds from local businessmen, although many struggled to raise such funds. The Wright brothers were the exception; their early publicity attracted third-party funding. The ability to attract non-family investment for early companies largely came from those in the broader aviation community, e.g. local businessmen with an interest in flight and competition wins. Aviation entrepreneurs who were successfully running businesses in related industries often cross-subsidised their flight business. Traditional business interests tended to view the floating of aviation companies as being especially risky business and it was not until much later that this type of finance raising became more common. Usually, there was an insufficient volume of investment from the public despite the general love of aviation the general public exhibited. Later, as the industry matured, business syndicates with an eye on profit put up funding, with this being particularly true in the United States. Coupled with the aforementioned tendency of the pioneers to be the only ones with aviation knowledge and expertise, this tended to encourage the earlier pioneers to be the ones that launched the initial enterprises in the industry.

12.3 Insights into entrepreneurial learning and knowledge acquisition

In emerging high-tech industries, such as the early years of aircraft manufacturing, there are very few chances for nascent entrepreneurs to find employment in established organisations in the emerging industry because of the industry's newness. Technology knowledge and skills, therefore, were developed from early education and work experience (particularly apprenticeships) in affiliated industries, such as locomotive or motorcar or manufacturing, which demanded similar skill sets. Apprenticeships and technical schools often provided an important substitute environment for the pioneers to hone their skills, becoming an important part of their preparation for entry into the industry. Personal hobbies and other interests also encouraged the pioneers/entrepreneur to apply these skills to developing flying models and often later to full-scale prototypes. Flying clubs created by the pioneers/entrepreneurs provided those interested in flying the opportunity to demonstrate, promote and share cutting edge knowledge. Learning in the chaotic environment during the early stages of the aviation industry was therefore enabled by an emerging aviation community of practice, which drew together interested practitioners from related industries and hobby groups who were used to actively learning together and co-creating relevant new knowledge. Air shows, exhibitions and early membership of key aviation clubs (e.g. in England the Royal Aero Club and Royal Aeronautical Society) acted as magnets, connecting these like-minded individuals with the resources and people of influence needed to deliver solutions and product. The flying clubs were also able to provide a forum which helped to identify opportunities and provide support for handling aspects associated with technical newness. Encouragement came from newspaper owners on the lookout for bold and exciting events. Their competitions offered important media attention and substantial financial prizes. Early winners of these competitions often made use of this prize money (and media attention) to develop new and tweak existing prototype machines and promoted these new machines. It could, therefore, be argued that the media industry (and to a lesser extent the air clubs) acted as a key enabler, collecting the many small public contributions (ticket entrance fees to flying shows and purchase of materials promoting these events) into a larger whole which could be used to fund prizes that contributed to supporting the growth of the industry.

Challenges were overcome and the needs of the emerging industry were met largely through the adoption of an approach favouring action over planning; experimentation and learning; and the conversion of experience to knowledge in an exploratory community of practice enabled way. The transitory and spiral careers in related fields of these engineering entrepreneurs allowed them to seek out work positions, activities and others who would allow them to pursue their love of aviation under their own terms. This richness in experience fed into the emerging aviation community of practice, bringing variety, independence and essential accumulative learning. The entrepreneurial need for creativity, learning and personal development stimulated the development of new knowledge sets. In the early stages of the industry, this was particularly important when knowledge had not yet been formalised.

Highly motivated individuals engaging at the inception of an emerging industry took on early leadership roles in the emerging industry. Those like the Wrights and Curtiss, who sought to control the industry by locking up technology knowledge through patents, often lost out – swept aside by governments, legislation and other entrepreneurs motivated to realise the potentials that aviation promised. The early entrepreneurs mainly acquired management experience as they worked in the industry, or alternatively brought in trusted others (e.g. family or fellow flying club acquaintances) to handle the financial and other non-technical and design aspects of their business. Communities of practice supported the emerging process, providing forums for co-learning as the industry lifecycle unfolded.

In the early stages of the aviation industry, prior experience in a related or emerging industry and connection to a vibrant communities of practice removed the need to complete schooling or apprenticeships in related fields. As the industry began to mature, the number of successful entrepreneurial companies in the marketplace expanded, offering the next wave of entrepreneurs the opportunity to be involved in and learn from early movers. Traditional educational institutions like universities caught up and began to establish formalised educational programs that captured and built on early learning. These, in turn, supported the building of networks for profit and influence that support the emergence of linear career paths. Therefore, to understand entrepreneurial learning and career development and entrepreneurial success in emerging industries, it is necessary to incorporate related industry experience and its contribution to the development of and learning within communities of practice.

12.3 Company structures

Most of the early pioneers struggled to raise capital. Early finance mainly came from other businesses held by the entrepreneurs in related industries, prize money, flying schools, private individuals and the sales of aircraft. Often, the firms once established used the finance they had managed to raise or borrow from personally owned related businesses to develop experimental aircraft. Then after a few sales of aircraft, a public issue of capital would be made, which would then enable a greater production capacity to be created. This was true in the United States, the United Kingdom and France, where all markets were affected by the build-up and drop off of demand after the world wars. A good example of this is the Hawker and de Havilland companies that in the 1920s had developed respectively the Moth and Hart range of aircraft. After initial sales were good (by Hawker to the military and de Havilland to commercial airlines), the companies raised capital through stock issues (in 1933 and 1928, respectively), which allowed them to expand their production facilities.

This process of development meant that the design teams' creative abilities were critical to the possibility of success. For example, the Wright brothers' attempt to control the early marketplace through patents without continuing to substantially innovate meant that by 1912, their aviation machines were considered inferior to those coming out of Europe. Similarly, in 1915, after a five-year fight with the Wright brothers over patents, Blériot was unable to secure orders for his plane by the British, who thought his plane was outdated. This suggests that design was at the heart of ongoing success. Well-designed aircraft able to meet the demands of commercial and military consumers created sales that allowed finances to be raised. This helped to transform the very nature of the companies themselves, from small design/workshop style companies into large-scale industrial enterprises. Often this transformation made it difficult for the initial pioneers to retain control of their companies, as the sheer scale of the operations taxed their managerial abilities. At this point a number of entrepreneurs left earlier start-ups, now maturing into larger organisations, to begin with new start-ups and new partners offering similar products.

The problem they faced was that the aviation entrepreneurs, more often than not, had little managerial experience. Contemporary research (e.g. Mueller, Volery, and von Siemens, 2012) suggests the behaviour of entrepreneurs in their ventures changes between start-up and growth. The growth of the venture moves business activities from open-ended, direct first-hand experience to a more closed, hands-off experience. Open-ended activities include defining an opportunity, developing a business plan, building a prototype and obtaining the first customer – all of which demand significant chunks of time and the capacity to scan the environment for resources. As the venture grows, the pace of work increases, the business expands and entrepreneurs have to coordinate more subordinates, production and distribution activities. First-hand, direct activities are replaced with managerial ones relating to growth in sales, market share growth and internal organisational mechanisms. Like entrepreneurs in dynamic markets today, entrepreneurs operating across the world war periods had to deal with factors still common now. Under dynamic conditions, the entrepreneur's business becomes more complex as they seek to balance demand for products, staff and the need to innovate to remain competitive. New products must be added to the portfolio and organisational structure needs to be adapted to respond to internal (e.g. growing or falling employee numbers, span of control) and external constraints (e.g. collapsing markets, meeting customer demand, entering a new market). Activities associated with growth (or rapid shrinkage) therefore take founders away from design and building work – the elements that stimulated and motivate them. In some cases (the Wright brothers, A.V. Roe, the Short brothers, Handley Page and Geoffrey de Havilland), family members were drawn into their enterprises to take on much of the responsibilities of conventional management. In other cases, such as the case of Glenn Curtis, the initial pioneers stepped back from overall management of the greatly enlarged companies and instead focused on design issues or eventually dropped out altogether.

In seeing the way in which the aviators/entrepreneurs began and operated their respective businesses, consideration of the contextual environment is important to understanding how they negotiated their organisations through this dynamic environment. Contemporary researchers often acknowledged that high-tech industries undergoing rapid technological change attract entrepreneurial types keen to develop new products and enterprises. The aviation pioneers tended to be people of this type.

The existence of entrepreneurial behaviour and the success of entrepreneurs does not only depend on the context of the industry in which they operate, but it also depends on the level of dynamism within the industry concerned. A high-tech industry undergoing rapid technological change alters the very conditions of the industry, changing the nature of entrepreneurial activity and opportunities available. When this change is driven by contextual needs, such as during World Wars I and II, the opportunities can emerge and disappear swiftly, which demands an agility and flexibility in entrepreneurs. Being connected to the right people in such a way as to be able to access these opportunities can be important. The Wrights, A.V. Roe and Thomas Sopwith, for example, started their first enterprises before World War I when demand for technology to create competitive advantage in battle were being sought. In contrast, McDonnell, Douglas and de Havilland's first firms were formed after World War I when the demand for military aircraft was falling away and when those in the industry were struggling to reposition themselves into alternative markets. These latter entrepreneurs tried, with some success, to develop commercial passenger aircraft, which were technically developing very rapidly at this time.

The fermentation periods of the aeronautical industry provide good examples of periods of very dynamic technological advances and large numbers of active entrepreneurs. The industry attracted and then gave entrepreneurial types the opportunity to establish new enterprises and to develop new and improved products when they were still at a relatively youthful age. Industry context, therefore, is important to the activities of entrepreneurs.

The aviation industry was an especially open one to new entrants in its earliest days because of the technological dynamism of it. The industry, however, did change over time due to two factors: firstly, an increasing complexity in design and manufacturing required additional amounts of diverse skills and expertise, and, secondly, action by early entrepreneurs to control the development of the market. The original model of a company with a design office and a large workshop based on an aerodrome was increasingly replaced with large-scale mass-production techniques. This meant that eventually in all three countries, the number of aircraft manufacturing companies declined, and companies were either merged or acquired by rivals. This process did not take place quickly and often the original pioneers were important in delaying this process. It is of note that major mergers in the industry, such as that of the Curtiss and Wright organisations, took place after the founders of the companies had retired or lost control of their companies. Others, like Blackburn and Handley Page, had more staying power,

Figure 12.1 Blackburn Buccaneer S2 production at Hawker Siddeley Aviation Brough

struggling to keep their companies independent but, eventually, these companies were subsumed into larger conglomerates after their retirement (see Figure 12.1). Many of these conglomerates were owned by entrepreneurs who sought to carve out and defend sections of the market through patents and contracts with the government. Eventually, of course, these pioneers fell by the wayside as more professional managers took over the running of the companies; nonetheless, some such as Sopwith, Martin, Douglas and McDonnell had great staying power, retaining important roles in the industry long into the post-World War II age of jet aircraft.

12.4 Conclusion

To be successful, aviation entrepreneurs had to be both starters and runners. In the periods of growth just prior to the world wars, government money allowed the starters to thrive, as available resources supported minor mistakes and adjustment. Once the world wars had passed, the runners became more important as efficient and strategically run organisations were needed. Innovative flair was essential to adapt to rapidly expanding or dwindling aviation demand. In these depressed periods, individuals from wealthy backgrounds or those who had built up a war chest of good business connections, money and clear leadership in a market sector tended to perform better than those entering the market for the first time. Early movers set the standards that others had to follow, and a review of their organisation's longevity and domination of later markets attests to this success. The legacy of the pioneers was a very long one, with aircraft and companies lasting a very long time (see Figure 12.2).

Figure 12.2 Flybe Bombardier Dash-8–400 (G-JECL) takes off from Manchester Airport, England, 29 March 2009

References

Fiol, C. M. and Lyles, M. (1985). Organisational learning. *Academy of Management Review*, 10, pp. 803–813.

Huber, G. P. (1991). Organisational learning: The contributing processes and the literatures. *Organisation Science*, 2(1), pp. 1–147.

Mueller, S., Volery, T. and von Siemens, B. (2012). What do entrepreneurs actually do? An observational study of entrepreneurs' everyday behavior in the start-up and growth stages. *Entrepreneurship Theory and Practice*, 36(5), pp. 995–1017.

13 Conclusion

13.1 Introduction

This book set out to explore how certain entrepreneurs negotiated the dynamic emerging aviation industry to create or adapt innovations to meet the needs of their target markets. The selected portraits chosen for this book examine how these individuals and their firms fared, and whether the insights provided for this examination have relevance for entrepreneurial activity today.

The aerospace industry today is one of the most technologically advanced ones in the world, producing a range of highly complex products. The industry today is dominated by mainly large corporations that are part of an extensive network of interconnected companies, with the lead times in invention and design of new aircraft now running into decades. These interconnected companies often receive substantial funding and backing from their governments due to the strategic nature of their business.

Along with the concentration and economies of scale and scope in production, is a countervailing power on the consumer side. Most aerospace demand still comes from the military of each nation along with a limited number of commercial passenger companies. Although the market for corporate and private aircraft does exist, it has never been as important as was originally envisaged when the industry was first created in the first decade of the twentieth century. Instead, the military still plays a crucial role in the development of the industry, along with a substantial commercial passenger industry that, since World War II, has met and exceeded the vision of the industry's pioneer aviators and entrepreneurs.

The industry is still mainly located in the countries where it first arose – the United States, France, the United Kingdom, Germany, Russia, Japan and Italy – and these countries remain important aerospace production facilities. This means that the legacy of the founders of the industry has been a persistent one, reaching into the twenty-first century.

The founders of the industry are an unusual group of individuals, entrepreneurial in temperament and experimentation and adventurous in practice. Although it might seem a contradiction to seek similarities in such an intensely individualistic group of people, there does emerge from the study evidence of some general common characteristics. As individuals, they do warrant further study. Knowledge of their achievements and how they seized opportunities in a constantly changing industry environment would go some way towards building up our knowledge of what makes an entrepreneur. What is most compelling about them is the realisation that they created not only aircraft that are some of the most iconic products of the twentieth century but also large-scale companies with innovative design centres that contributed to much of the technological progress of the twentieth century. Examining shifts in the aviation industry and the type of entrepreneurial activities needed to address these shifts suggests patterns in behaviour that are applicable today. In the next section, we overview these shifts and responses in order to be able to discuss their relevance today.

13.2 Entrepreneurial engagement with context

As has already been discussed, from the 1900s to 1945, the aviation industry experienced periods of significant change as countries involved in its development moved through periods of peace, rearmament, wartime and mobilisation. Each period came with its own opportunities and challenges which favoured particular entrepreneurial activities such as developing a new product, introducing new processes such as in manufacturing, opening up new markets, finding new sources of supply for materials, creating competitive positions in the marketplace and innovating to allow technology transfer from one area to another.

13.3 Peacetime

Periods of peace existed at the start of the aviation industry; post-World War I (1924 to the early 1930s) and post-World War II (from the late 1940s to the early 1950s). During these periods, money was in short supply and government decision-making regarding aviation lacked urgency. There was no pressure to invest heavily in technology development and no wars in which armed forces could fully test equipment or doctrine. Orders for aircraft were slow to materialise and delivery dates stretched out. Political pressures often overruled the best designs. Existing designs were frequently modified to solve emerging problems or exploit emerging technology as part of the development and delivery process. For entrepreneurs operating at this time, this period was like hiking in a desert with limited water rations. Peacetime conditions needed flexible and inexpensive firm structures that could respond quickly to opportunities as they arose. It favoured design establishments with workshops attached – conditions which favoured entrepreneurs specialising in the development of new products, innovation and technology transfer from one area to another.

At the inception of the aviation industry, there was a lack of government interest and take up. In the interwar period of peace between World War I and World War II, government orders for military aircraft dried up quickly, leaving firms with a surplus of aircraft and no market for them. World War II military planes were ill-suited for the commercial and recreational market. In addition, the Depression years wiped out the domestic market for recreational aircraft, drying up income sources. A few British firms were reasonably successful at selling armaments overseas, but those without the entrepreneurial ability to source new markets for existing product struggled. Firms attempted to diversify into markets where their raw materials could be reused, such as the making of pots and pans, which required different production processes. This led to operational inefficiencies. The British Government's move to protect a key group of manufacturers by drip feeding them small military orders – mainly refurbishment of existing craft for military use – locked these companies into a military provision framework which worked against them adapting to the emerging commercial market. In America, entrepreneurs fared better because the government there saw the value in supporting the development of the commercial aviation market. Although in both America and the United Kingdom entrepreneurial activities focused on new product development, and the opening up and structuring of new markets were needed for entrepreneurial survival, contextual factors shaped what entrepreneurial activity was actually possible in each country.

13.4 Rearmament

Rearmament is often described as a state of confusion when governments must decide whether to invest time designing the most up to date aircraft or to go for quantity production with existing models. Money is less of an issue, but key limitations exist around the labour force, the supply of raw materials and the competitive demands of buyers and changing war contexts. Government departments have to decide what aircraft type is needed as each aircraft type needs specifications drawn up and bids sent out. Increasing aircraft types reduce the number of tenders submitted as non-standardisation of parts increases cost and production/operation problems for the manufacturers. Firms must decide which designs to tender for, i.e. which designs match their own ability and facilities, and which are likely to deliver future sales. Once designs are accepted, paper proposals must be converted to prototype and then to production models. The longer the lead time from prototype to production, the longer the period before real payback on investment begins. In World War I, aircraft designs were simple enough for several to be developed in one year, with prototypes being brought to production within 12 months. Planes were handcrafted, workshop-based affairs, which provided firms with greater flexibility which lowered the risk and cost of pursuing designs that were not subsequently taken up. After World War II, however, risk and cost rose as greater technology sophistication stretched the lag time from aircraft design to production to more than five years. Therefore, entrepreneurs operating during World War I faced a different risk profile to those operating during World War II.

Other factors that could raise operating costs by stretching the lead time from aircraft design to production were the involvement of multiple stakeholders with differing perspectives; insufficient technical instructions and tools; and scattered facilities and staff. Major delays could ultimately lead to the abandonment of designs and planes, particularly during wartime. For example,

> between 1934 and 1942 the RAF had fifty-six types [of planes] in production in the United Kingdom and nineteen more on order in the United States. Of these seventy-five designs, only eighteen went into mass production, and only eight of these types were produced in volumes more than 5,000 each.
> (Higham, 1968, p. 453)

Considerable pressure was therefore placed on entrepreneurs and their firms to choose wisely about where to innovate, what technology to transfer to new product development and what production processes to develop, ramp up and tweak. For World War II, the design-to-production phase had to absorb significant technological developments and be re-oriented to high-speed quantity production. This demanded longer and more sophisticated production lines to accommodate metal construction, high h.p. engines and pressurisation. Specialists and new processes, procedures and machinery were needed but these were hard to come by as they often had to be sourced from related industries (e.g. the motorcar industry), which were also under pressure to grow. Connections within the industry could help to facilitate access to these resources, as well as government support. For example, in the 1930s, to ease the pressure, the British Government restricted demands from some areas (e.g. the British Army) in order to concentrate upon air defence, allowing capacity from the motorcar industry to be used in aircraft production.

The shift from quality to quantity production also raised important strategic decisions for aviation manufacturers. In a complex, shifting, unpredictable market, firms needed to secure high profits wherever possible to protect themselves against premature cancellation of orders. They also needed to maintain a good financial record, reflected in dividend payments, to attract investors who were nervous about parliamentary investigations, government contract renegotiation or even actual state takeover of their companies. These three events were more prevalent during the rearmament period. In addition, the move from quality to quantity production initially increased accident rates as inexperienced technicians served inexperienced flying crews in radically more sophisticated aircraft whose inherent design faults began to show up under constant use. Manufacturers tended to keep the causes of accidents secret to avoid damaging sales or attracting lawsuits while simultaneously quickly learning and improving quality. These factors made firms reluctant to take on the capital debt required to create mass-production factories, so governments initially supported this transition through investment. In the United Kingdom, investment money was primarily to a chosen 'ring' of government-supported suppliers.

Rearmament confusion required the full range of entrepreneurial activities to allow firms to respond quickly to changing environments, i.e. new goods, new production methods, opening new markets, sourcing new materials, structuring industry sectors, innovating and technological transfer. In the United Kingdom, where about a fifth of the RAF budget was focused on prototype development, the focus was heavily on new goods and new production methods.

13.5 Wartime

In wartime, target delivery dates rather than money become a priority. Production problems and procedures are ironed out, leading to the steady flow of aircraft, engines spares and construction of new machines or repair of damaged ones. The availability of manufacturing facilities, raw materials and political preferences dictate the variety of designs and production types. Government and government organisations attempt to balance flexibility with quality and standardisation for quantity. Designs and production are usually assigned to a select group of manufacturers with group outsiders being invited to contribute during periods of crisis in wartime. Once the crisis passes, fresh designs are set aside in order to return to standardisation and quality. In wartime, flexible designs that can be mass-produced, used in a wide variety of operational settings and easily adapted to more power, heavier weights, armaments or fuel are popular. Projects unlikely to be completed in time are cut back or even abandoned, although knowledge arising from accidents and rapid technological developments may convert some unsuccessful designs into successful ones. Equipment unable to stand up to the enemy is relegated to operations, training or transport. The influx of non-professionals increases demands for change as intensive use reveals a greater need.

This pattern of behaviour was particularly evident in the United States and the United Kingdom during the world war periods. Entrepreneurial activity during this period was focused on enhancing production techniques and using resources effectively, though not necessarily efficiently.

13.6 Demobilisation

Demobilisation marks a period of instability as politicians and firms look to the future and peace. The focus shifts to developing products likely to be in demand and saleable when crisis/conflict ends. This period may occur even before a war starts, and in the case of governments, they can develop policy that shapes industry development in the post-war period.

For example, between 1919 and 1945, the British Government's lack of acceptance of the need for larger armed forces equipped with modern materials to protect national security led to a cut in expenditure on armaments projects. Equally the British Government was uninterested in commercial aircraft, despite committee advice (see the 1942 Brabazon Committee report on types of airliners likely to be needed after war). This policy decision led to the British cancelling commercial designs in progress in 1939 and in 1941, reaching an agreement with the Americans that the United Kingdom would not produce transport aircraft. Consequently, America established a lead in this area, while a gap in commercial flight designs and planes emerged in the post-war United Kingdom. When transport planes were eventually needed for British Overseas Airways Corporation (BOAC), the British ended up buying Canadian planes, allowing for quantity, quality and politics to be satisfied.

For firms, the demobilisation period means entrepreneurial activity shifts to positioning the firm for peacetime, from the present to the future. Entrepreneurial activities move to carve out a market position, through new product, new processes or new technology. Time ceases to be of the essence – manufacturing slows and quality slips. Management reallocates personnel to projects promising better returns than military production. Employees drift off to other industries. Companies begin to go back to a pre-war structure – often incorporating simple organisational structures consisting of design establishments with attached workshops.

Post-World War I, the British aviation industry was dominated by those who had been in charge when war started or who were industry pioneers. Post-World War II, dramatic technological developments meant a skilful workforce needed to be retained, and production runs needed to be maintained to protect capital investment. However, experimenting with new craft demanded smaller production runs, which proved a challenge for many companies.

13.7 Popular times to start a business

Most of the entrepreneurs studied in this book started their companies, particularly their first companies, during periods of peace, i.e. pre-World War I or during the interwar years. This suggests a preference for the entrepreneurs studied for innovation and technology transfer and new market development. This favoured smaller, design-type organisational structures. Those starting companies in the interwar years and post-World War II periods did so in an attempt to diversify into other related markets for survival or to capture promising new markets like commercial airlines. These company formations were predominantly financed by private and public investment, as little government funding was available.

The second most popular time to start a company was during armament when military money was flowing into the market – this was particularly evident amongst the American entrepreneurs studied. Rearmament demands a wider range of entrepreneurial activity and skill sets beyond new market development and the innovation and technological transfer activities needed in peacetime. Rearmament also brings different risks such as the potential acquisition of key firms by governments (such as in France and the United Kingdom). Acquisition of firms by a government often marked a point of exit from the industry for the founders, but some did go on to later found new companies.

Of the 25 early aviation entrepreneurs studied, only Wright and Curtiss exited from the industry pre-World War II. Others like Lockheed, Martin, Saulnier, the Short brothers, Handley Page, Sopwith, McDonnell and Douglas were absorbed and merged with other aviation companies into major corporations still operating today and contributing to the American, British and French economies (see the Appendix). For example, in the United Kingdom, BAE Systems now encompasses companies founded by Sopwith, A.V Roe, de Havilland, Shute Norway and Handley Page. In America, Curtiss-Wright, Lockheed Martin and Boeing still exist – McDonnell Douglas ceased to exist in the late 1990s. In France, Dassault Aviation still exists.

13.8 Surviving the phases of peacetime, rearmament, war and demobilisation

For entrepreneurs and their firms to survive across the phases of peacetime, rearmament, war and demobilisation required significant adaptability as the demand for particular types of entrepreneurial activity shifted. These entrepreneurial activity types incorporated developing new products; introducing new processes, such as in manufacturing; opening up new markets; finding new sources of supply for materials; creating competitive positions in the marketplace; and innovating to allow technology transfer from one area to another. Early peacetime conditions required flexible and inexpensive firm structures that could respond quickly to seize opportunities. Rearmament flooded the market with money focused on particular military outcomes, making the market more forgiving of weak strategy and organisational structures. However, military outcomes often took those entrepreneurs stimulated by innovation and design away from their centre. War demanded primarily efficiency, focus and standardised designs. Demobilisation was perhaps the period of greatest risk as the standardised world of the mass-production market dried up, and there was a return to experimentation and innovation as new industries were formed. Failure of entrepreneurs and their firms to respond strategically and structurally to market changes could, therefore, be partly explained by their inability to shift between innovation and creativity and standardised production.

13.9 Entrepreneurs' use of communities of practice

The use of communities of practice by the aviation community played an important role in developing the aircraft industry. These communities of practice, which at their broadest level incorporated media, flying clubs, governments, trade organisations and philanthropists, were essential in building the legitimacy of the technology in the government and public's eyes. The technical community of practice was particularly useful for entrepreneurs in periods of significant development when the boundaries of design and production were being extended or in times of industry contraction when resources and ideas came primarily from within the aviation community itself. During these periods, existing knowledge was insufficient to meet current needs as the changing technological dimensions of the industry curtailed the transferability of previous knowledge. Previous discussion has shown the mixed benefit of government intervention and university support due to competing priorities. Both of these groups were slow to engage with aviation, largely because they did not fully understand its value and relevance. When the American and British Governments did engage, their focus was on military rather than commercial interests, although the Americans, due to America's larger size, were quicker to realise the ability of air to efficiently connect their states together. Both governments established research institutions in the 1900s – the United Kingdom's Advisory Committee for Aeronautics in 1909 and the United States' National Advisory Committee for Aeronautics in 1915. Both institutions focused on resolving key aviation research problems, primarily of a military nature, although America's research institute had a wider remit, covering both military and later commercial considerations, which allowed America to provide testing facilities to private manufacturers earlier than the United Kingdom.

In the education space, the skill sets needed to develop the aviation industry came from technical school training or on the job training in related fields. Larger institutions like universities offering aeronautical degrees did not emerge until the early 1920s. Their focus was on engineering rather than on business aspects associated with running an aviation firm. This meant that at key developmental or contraction stages in the industry, entrepreneurs needed to work with their community of practice to resolve issues.

In the early stages of the industry, aviation clubs and the media were important parts of the network supporting and promoting aviation technical and business development. For example, in the early stages, the Wright brothers shared their aviation experiments through the aviation clubs of Europe, stimulating the drive there to establish aviation. Aviation clubs were also important contributors to funding early development at a time when media in America, the United Kingdom and France were as sceptical as the governments about the capabilities of flight.

Once volume production of a small range of designs emerged, the aviation industry shifted to mass-production techniques. In the United Kingdom, in the interwar years, some firms sought to adopt mass-production techniques in order to compete with America. For example, Airco visited America to study the mass-production methods for motorcar production in a bid to diversify. Although Airco subsequently collapsed, its assets were sold to de Havilland with backing from Airco's original founder, George Holt Thomas. Knowledge of motorcar mass production positioned the new de Havilland Company well for aviation mass production.

13.10 What this means for entrepreneurship today

A key question is what relevant the findings in this book have for entrepreneurs and their firms today. To answer this question, we look at the rise of the drone. A drone is an unmanned aerial vehicle (UAV) with a certain degree of autonomy (Giones and Brem, 2017). Recent developments in the range of applications of drones has turned the drone industry into a multi-billion industry with both military and commercial applications. The development of the drone industry has many similarities with the development of the aviation industry. This suggests that comparing and contrasting aviation and drone industry development could yield useful insights for entrepreneurs and their firms.

Drones were first developed towards the end of World War I, but it was not until World War II that they became extensively used in war, by the Germans as drone bombs. Since then military drones have been extensively used for reconnaissance missions (1950s and 1960s), as decoys (1970s), for real-time surveillance and radar (1980s and 1990s), for combat and surveillance and to carry armaments (2000s) and for combat and attacking targets (2010 onwards). The civilian drone market started to develop in the mid-1990s where they were used for mapping, aerial photography and logistics. In the 2000s, drones moved into industrial inspections, security and precision agriculture. In the 2010s, drones have been used extensively in filming and broadcasting and for recreational use (Giones and Brem, 2017).

Similarly, patterns in development occurred in aviation with the military using planes firstly for reconnaissance missions, then for surveillance and radar and finally for combat. The key difference is in the area of civilian engagement, which was evident at the start of the industry (recreational flyers); in the interwar years (filming and joy rides); and post-World War I (commercial and passenger transport) (Giones and Brem, 2017).

Like aeroplanes a century before, for a long period of time drones remained on the margins of warfare due to technical limitations (e.g. a lack of endurance in the air) and vision about their domestic and military capability. Early aeroplanes and early drones were initially viewed by the military as being insufficiently robust for full operations and were used for reconnaissance. It was only after major events (for aviation, World War I; for US manufacturers of drones, September 11) that there was a rapid increase in military interest. Once the military potential was identified, military money supported further development and growth.

Both aeroplanes and drones emerged as a result of combining a number of technologies: aerial capacity, flight control, position control and communication, and multiple actors with a shared agreement to develop and exploit the opportunity. In both industries, talented innovators, attracted to something exciting and cool, had to work with others from related to technologies to overcome technology constraints which could not be solved from the knowledge base of one industry. This meant that the development of early aviation and early drones emerged from small size businesses – cottage based (aviation) and garage based (drone). These small-sized businesses were funded initially through competition winnings, small grants, personal money and a handful of early believers. Once the military came on board, they funded development in areas of military significance. In both areas, the development of major non-military interests came later in both industries' development – commercial flight in aviation, civilian applications in drones (Giones and Brem, 2017).

A key difference between the two is the cost of production. In aviation as plane technology improved, the cost of production increased dramatically, moving production to large-scale developers. By comparison, generally, the cost of drones has fallen over time (except for highly specialised military ones requiring sophisticated technology). The comparatively low cost of drones compared with aviation has facilitated the development of an increasingly important civilian market for drones. This civilian market has extensively used communities of practice to co-create, test and develop new technologies (Giones and Brem, 2017).

Differences also exist in the lifecycle periods of the industries. The aviation industry went from emergence through growth to maturity in 50 years – the drone industry has taken over 100 years to reach the early stages of major rapid growth. It will be interesting to see if once growth peaks in the drone industry whether maturity comes quickly and innovation falls off, as was the case in the aviation industry.

Both industries have experienced cycles of emergence and contraction but the magnitude of these to date appear to have been less in the drone industry. This may be due to the lower cost of experimentation and equipment compared with planes, meaning the benefits of drones outweigh the cost in high-risk situations and promise significant cost savings in commercial operations. However, high growth periods in both industries have led to design flaws and the malfunction of technical components resulting in the jeopardising of key aviation and drone companies. Currently, the drone industry appears to be moving into a period of mergers and acquisitions as happened in the aviation industry after World War I and World War II. If patterns from the aviation patterns are repeated in the drone industry, then it is likely that the early mover companies will be subsumed by second-generation companies but their DNA will remain as key building blocks of those organisations will survive into the future. It can already be seen that second-generation drone companies comprise both aviation and non-aviation enterprises, e.g. Boeing, Google and Amazon. This suggests that in the drone industry there is both consolidation and stabilisation occurring at the same time as growth in new applications. Given the overlap between the two industries, it will be interesting to see if patterns that appeared in the maturing of the aviation industry also play out in the drone industry. Could passenger transport be the future for drone technology? (Finn, 2012).

Studying both industries suggests that entrepreneurs and their firms need similar adaptability and flexibility to survive.

13.11 Conclusion

Every industry has its own patterns of growth and contraction which require different types of entrepreneurial engagement. The entrepreneur profiles studied suggest different entrepreneurs show preferences for particular entrepreneurial activities and that these preferences can make it difficult for them to adapt to environments demanding a different entrepreneurial activity profile. In industries where change is frequent and rapid, the pressure on entrepreneurs to adapt can, therefore, be intense and stressful. External pressures like war add additional uncertainty. Shifts in government policy and funding can ease or exacerbate this pressure by aiding or retarding industry ebbs and flows, which in turn can have an effect on the ability of entrepreneurs to transition across change phases. Under these conditions, it is the entrepreneurial pursuit of a vision that allows many to fail forward.

This study also suggests that communities of practice are particularly useful during periods of dynamic change when established governments and educational institutions are unable to keep up. During rapid expansion and contraction periods, it is those embedded in the area who are best positioned to encourage and fund experimentation and start-ups, suggesting they could provide an important bridge between institutions (e.g. governments) wishing to invest wisely. As the technology and industry stabilises, the relevance of governments and educational institutions increases as they provide links to a broader ecosystem encompassing research institutions, start-ups, investors and government funding necessary for embedding and building on the changes and for giving the technology credibility. Finally, governments and research institutions must seek to structure a market that rewards productive entrepreneurial ability that correctly identifies and responds to the needs of the marketplace.

References

Finn, P. (2012). Rise of the drone: From California garage to multibillion-dollar defense industry. The *Washington Post*, 23 Dec.

Giones, F. and Brem, A. (2017). From toys to tools: The co-evolution of technological and entrepreneurial developments in the drone industry. *Business Horizons*, 60(6), pp. 875–884.

Higham, R. (1968). Quantity vs. quality: The impact of changing demand on the British aircraft industry 1900–1960. *Business History Review*, 42(4), pp. 443–466.

Appendix

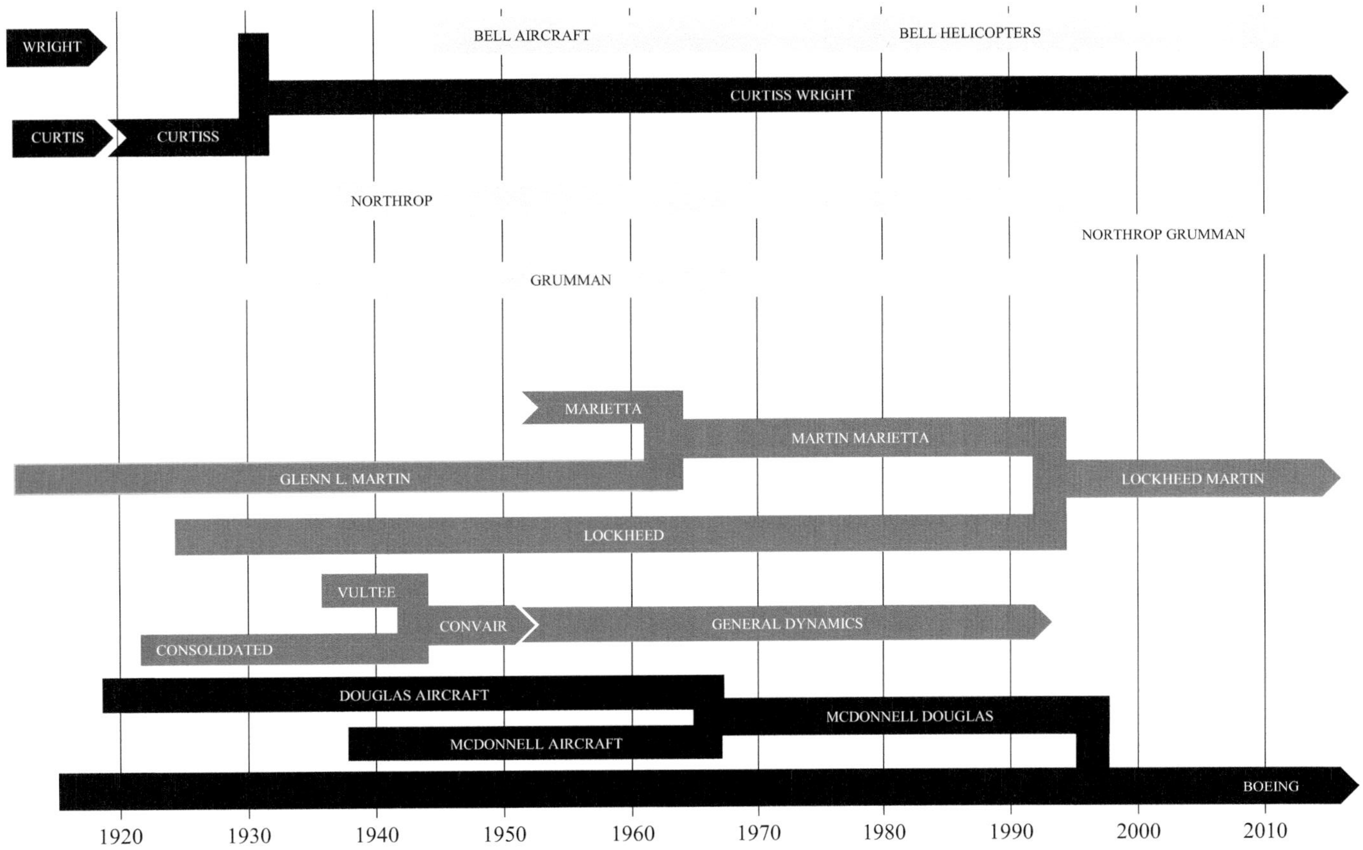

Figure A.1 Major United States manufacturers of aircraft, 1908 to 2018

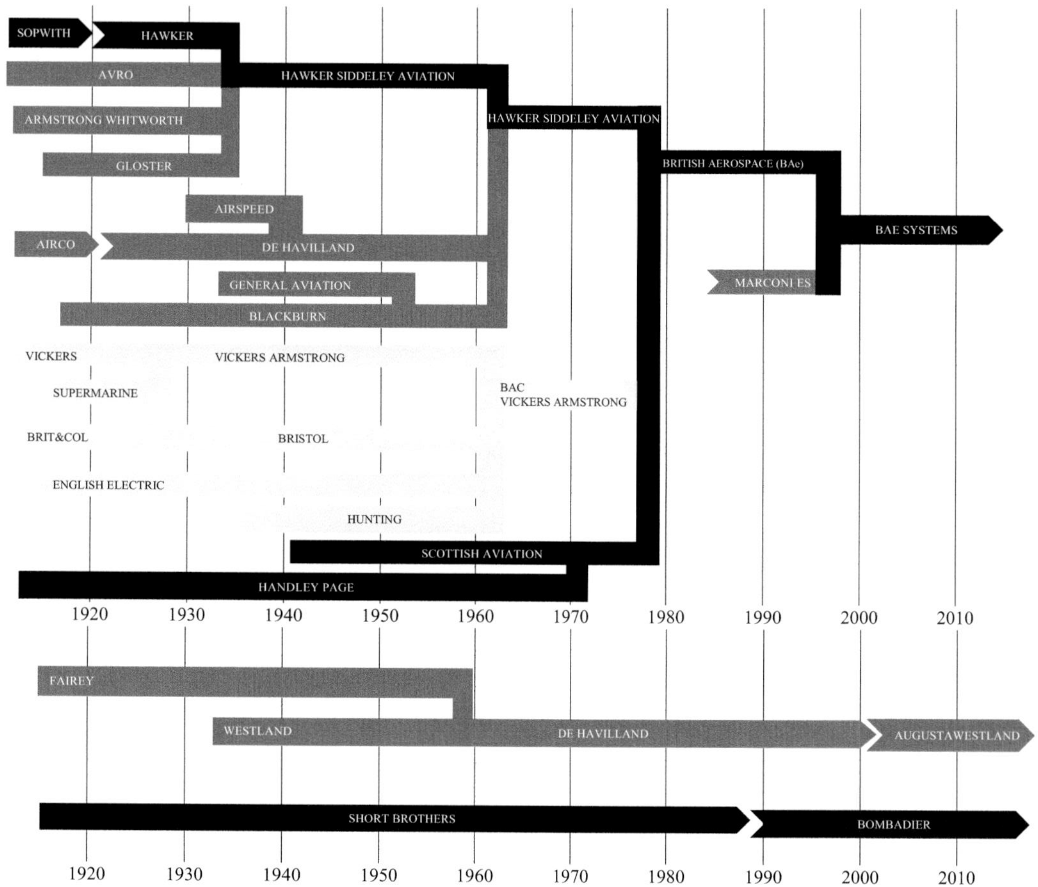

Figure A.2 Major British manufacturers of aircraft, 1908 to 2018

Table A.1 Prominent aviation pioneers/entrepreneurs in the United Kingdom, the United States and France

Name	*Life*	*Study field*	*Schooling*	*Early work*	*Hobbies/ interests*	*Company/ companies*	*Aero club licence*	*FRAes*
United Kingdom								
Frederick Handley Page	1885–1962	Electrical engineering	Cheltenham Grammar; Finsbury Technical College	Electrical engineering	NA	Handley Page Ltd.		Yes
Geoffrey de Havilland	1882–1865	Mechanical engineering	Noneation Gramma; St Davids School; Crystal Palace School of Engineering	Apprenticed at steam engine manufacturers, draughtsman at tool and motorcar company, design office motor omnibus construction company	NA	De Havilland Aircraft Co. Ltd.	UK 53	Yes
Montague Napier	1870–1931	No formalised study past school	Secondary school	Took over family engineering business, cars and diversified into tools for the cycle industry and aircraft engines	Motorcar racing	D. Napier & Son Ltd.		
Frederick Henry Royce	1863–1933	Electrical engineering	Elementary school	Apprenticed at Railway Company locomotive works	Motorcar racing	Rolls-Royce Ltd.		
George Holt Thomas	1869–1929	Incomplete degree	King's College School; Queen's College Oxford	Director of newspaper company		Aircraft Manufacturing Co. Ltd. (Airco)		

(*Continued*)

Table A.1 (Continued)

Name	*Life*	*Study field*	*Schooling*	*Early work*	*Hobbies/ interests*	*Company/ companies*	*Aero club licence*	*FRAes*
Alliot Vernon Roe	1877–1958	Marine engineering	St Paul's School London; Kings College London	Apprenticed at locomotive works, motorcars	Bikes, flying models	A.V. Roe & Company (Avro), Saunders-Roe	UK 18	Yes
Thomas Sopwith	1888–1989	Engineering	Cottesmore School Sussex; Seafield Park Engineering College	In business with others selling motorcars	Ice skating, ballooning, motorcycles, car racing, sailing	Sopwith Aviation Co., Hawker Aircraft, Hawker Siddeley	UK 31	Yes
Charles Richard Fairey	1887–1956	Electrical engineering and chemistry	Merchant Taylor's School; Ardingley College; Finsbury Technical College	Worked in company manufacturer or arc lamps – responsible for installing lights	Skilled designer and builder of model aeroplanes and active member of the Kite and Model Aeroplane Association	Fairey Aviation Co. Ltd.		Yes
Oswald Short	1883–1969	Engineering	Secondary school	Office boy then owned balloon company	Ballooning	Short Brothers PLC		Yes
Eustace Short	1875–1932	Engineering	Secondary school	Owned balloon company	Ballooning	Short Brothers PLC		
Robert Blackburn	1885–1955	Engineering	Leeds Modern School; University of Leeds	Worked as manager at lawnmower and steamroller manufacturing	Motorcars, golf	Blackburn Aircraft Limited		Yes

George White	1854–1916	Apprenticeship junior clerk solicitors	St. Michaels Boys School	Owned stockbroking business; tram business, aeroplane business	Transport, later MP	British and Colonial Aeroplane Co.		
Nevil Shute Norway	1899–1960	Engineering	Shrewsbury School; Balliol College Oxford	Internships at Airco and de Havilland Aircraft Company in design office	Flying, motorcar racing, writing	Airspeed Ltd.		
Sam Saunders	1857–1933	NA		NA	NA	S.E. Saunders (Saunders-Roe)		
Noel Pemberton-Billing	1881–1948	Law	Elementary school	Ran away from home at 13 – mounted police, boxer, selling steam yachts, owned flying field	Own garage then moved to found aviation field, later MP advocating air force	Pemberton-Billing Ltd. (Supermarine Aviation Works Ltd.)	UK 632	
Ernest Petters	1873–1954	NA		Petters Engineering company	Built horseless carriage, unsuccessful MP	Westland Aircraft Ltd.		
Hugh Burroughes	1884–1985	NA		NA	NA	NA		
John Davenport Siddeley	1866–1953	Draughtsman apprentice	Secondary school	Draughtsman for cycle company; Dunlop's sales manager; manager Wolseley tool and motorcar company	Bicycle racing, cars	Armstrong Siddeley		
James Martin	1893–1981			Established own engineering company		Martin Baker Aircraft		Yes

(*Continued*)

Table A.1 (Continued)

Name	*Life*	*Study field*	*Schooling*	*Early work*	*Hobbies/ interests*	*Company/ companies*	*Aero club licence*	*FRAes*
United States								
Wilbur Wright	1867–1912	Self-taught	High school	Editor, bicycle retailer/ manufacturer, aircraft inventor/ manufacturer, pilot trainer	Ice skating, flying, bicycling	Wright Co. 1909–1929	US 5 honorary	
Orville Wright	1871–1948	Self-taught	High school	Editor, bicycle retailer/ manufacturer, aircraft inventor/ manufacturer, pilot trainer	Flying, bicycling	Wright Co. 1909–1929	US 4 honorary	
Glenn H. Curtiss	1878–1930	Interest mechanics and inventions	High school	First job at film company	bicycle racer and builder, motorcycles, aviator	Curtiss Aeroplane and Motor Co.	US 1	
Glenn L. Martin	1886–1955	Business	Kansa Wesleyan College	Several plane businesses	Kites	Glenn L. Martin Co., Wright-Martin		
Frederick B. Rentschler	1887–1956	Engineering	Princeton University	Ran family moulding and machinist business, later president of Wright Aeronautical Corp.	NA	Pratt & Whitney Aircraft Co.		
Allan Loughead	1889–1969	Self-taught	Elementary School	Mechanic, driving racing cars	Kites, gliding, driving racing cars	Alco Hydro-Aeroplane Co., Loughead Aircraft Manufacturing Co.		

William E. Boeing	1881–1956	Dropped out of Yale to work	Yale University	President of Greenwood Timber Co.	Flying, property development, horse breeding	Boeing Co.
Jack Northrop	1895–1981	Draughtsman	Secondary school	Draughtsman for Lockheed	Flying, tailless flight	Avion Corp., Northrop Corp.
Donald Wills Douglas	1892–1981	Aeronautical engineering	Naval Academy; MIT	Naval Academy, Connecticut Aircraft Co. (designer), US Army Signal Corps (civilian aeronautical engineer)	Model planes	Douglas Aircraft Co.
Larry Bell	1894–1956	Mechanic	Secondary school	Shop foreman at Glenn L Martin Co. then VP and GM at Consolidated Aircraft	NA	Bell Aircraft
Gerard Freebairn Vultee	1900–1938	Aviation science	Secondary school; evening classes at Cal Tech	NA	NA	Vultee
Vance Breese	1904–1973	Various engineering extension courses	Secondary school; evening classes	NA	NA	Breese Aircraft Co., Vultee

(*Continued*)

Table A.1 (Continued)

Name	*Life*	*Study field*	*Schooling*	*Early work*	*Hobbies/ interests*	*Company/ companies*	*Aero club licence*	*FRAes*
Sherman Fairchild	1896–1971	Dropped out several universities due to ill health	Harvard University	Entrepreneur (family money)	Inventor of synchronised camera shutter and flash, pioneer aerial photographer and filmer, studied architecture, cooking, jazz, dancing, philosophy and tennis	Fairchild Aircraft		
James Work	NA	NA		NA	NA	Brewster Aeronautical Corp.		
James Smith McDonnell	1899–1980	Aeronautical engineering	Princeton MIT	Ford Motor Company Airplane Division, Huff-Daland Airplane Company	NA	McDonnell Aircraft Corp.		
Reuben Hollis Fleet	1887–1975	Dropped out several universities	Culver Military Academy; Stanford university	Teacher, realtor, national guard, real estate	Later set up San Diego Air and Space Museum, Bishop Centre for Performing Arts	Consolidated Aircraft		

James Kindelberger	1895–1962	School dropout	Secondary school	Engineer at Douglas Aircraft	NA	North American Aviation
Leroy Grumman	1895–1982	Mechanical engineering	Cornell University	New York telephone company, US Reserve	NA	Grumman Aeronautical Engineering
Chance M. Vought	1890–1930	Engineering	New York University, University of Pennsylvania	NA	NA	Lewis and Vought Corp., Chance Vought Corp.
Alexander P. de Seversky	1894–1974	Engineering – then aeronautics	Imperial Russian Naval Academy Military School of Aeronautics	Military service, restaurant operator	Taught to fly by father at young age	Seversky Aero Corp.
Grover Cleveland Loening	1888–1976	Aeronautical engineering	Columbia University	Managed the Wright Company factory	Published book on military planes	Loening Aeronautical Engineering Corp., Grover Loening Aircraft Co.
Thomas Huff	NA	NA		NA	NA	Huff-Daland Aero Corp., Huff-Daland Aero Co., Keystone Aircraft Corp.
Elliot Daland	NA	NA		NA	NA	Huff-Daland Aero Corp., Huff-Daland Aero Co., Keystone Aircraft Corp.

(Continued)

Table A.1 (Continued)

Name	*Life*	*Study field*	*Schooling*	*Early work*	*Hobbies/ interests*	*Company/ companies*	*Aero club licence*	*FRAes*
Giuseppe Mario Bellanca	1886–1960	Engineering	Politecnico di Milano (Italy)	Operated the Bellanca Flying School	NA	Bellanca Aircraft Co.		
FRANCE								
Gabriel Voisin	1880–1973	Industrial design		Architectural firm and then worked for Ernest Archdeacon	Motorcars and motorbikes (design and building)	Voisin, Société Anonyme des Aéroplanes G. Voisin		
Charles Voisin	1882–1912	NA		NA	NA	Voisin		
Henri Farman	1874–1958	Painter	École des Beaux Arts	Own flying school	Painter, new mechanical inventions, champion cyclist, motor racing	Farman Aviation Works		
Louis Blériot	1872–1936	Science		Electrical engineering company	First practical headlight for motorcars	Blériot Aéronautique		
Robert Esnault-Pelterie	1881–1957	Engineering		NA	Designed flight joystick	REP		
Lois Breguet	1880–1955	Electrical engineering	Institute Notre Dame École Centrale	Developed gyroplane under Charles Richet	Founded organisation which became Air France	Breguet Aviation		
Edouard de Nieuport	1875–1911	NA		NA	Died young but aircraft designs won many	Société Anonyme des Établissements Nieuport		

					competitions and awards; company run by brother	
Louis Bechereau	1880–1970	NA		Mechanical construction firm	Model making	Société de Production des Aéroplanes Deperdussin
Raymond Saulnier	1881–1964	Aeronautical engineering		Collaborated Blériot on the channel crossing flight and then formed company with Moraine brothers	NA	Morane-Saulnier
Aldophe Bernard	1881–1955	NA		NA	NA	Avions Bernard
Émile Dewoitine	1879–1979	NA		Worked first for Groupe Latécoère	NA	Constructions Aéronautiques Émile Dewoitine
Felix Amiot	1894–1974	NA		Grew up in Issy-Les-Moulineaux home of Blériot, Voisin, Farman aircraft test sheds, recruited from Army to work in Moraine-Saulnier during World War I	Mechanics and aviation	Avions Amiot
Pierre-Georges Latécoère	1883–1943	Manufacturing		Fathers firm manufacturing railway wagons (modernised it)	Aeronautical enthusiast	Groupe Latécoère

(*Continued*)

Table A.1 (Continued)

Name	*Life*	*Study field*	*Schooling*	*Early work*	*Hobbies/ interests*	*Company/ companies*	*Aero club licence*	*FRAes*
Henry Potez	1891–1981	NA		NA	Inventor of propeller with Dassault	Aviation Potez; Lioré-et-Olivier, CAMS, Romano		
Marcel Bloch (Dassault)	1892–1986	Electrical engineering		Worked at French Aeronautics Research Laboratory during World War I	NA	Avions Marcel Bloch, Dassault aviation		

Index

Note: Page numbers in **bold** indicate tables; page numbers in *italic* indicate figures.